THE
LAYERED
EDIBLE
GARDEN

Quarto.com

© 2024 Quarto Publishing Group USA Inc.
Text © 2024 Fluent Garden Consulting

First Published in 2024 by Cool Springs Press, an imprint of The Quarto Group,
100 Cummings Center, Suite 265-D, Beverly, MA 01915, USA.
T (978) 282-9590 F (978) 283-2742

Cool Springs Press titles are also available at discount for retail, wholesale,
promotional, and bulk purchase. For details, contact the Special Sales Manager by
email at specialsales@quarto.com or by mail at The Quarto Group, Attn: Special
Sales Manager, 100 Cummings Center, Suite 265-D, Beverly, MA 01915, USA.

28 27 26 25 24   1 2 3 4 5

ISBN: 978-0-7603-8559-3

Digital edition published in 2024
eISBN: 978-0-7603-8560-9

Library of Congress Cataloging-in-Publication Data available.

Design: Cindy Samargia Laun
Photography: See page 215
Illustration: Erin Lowe @erindynamic

Printed in China

THE LAYERED EDIBLE GARDEN

A Beginner's Guide

to Creating a Productive

Food Garden Layer by Layer

Christina Chung of Fluent Garden

COOL
SPRINGS
PRESS

Layered edible gardens are created by combining plants with different statures, from trees and shrubs to ground covers. How many edible plants can you see in this foundation planting? There are cherry and apple trees in the tree layer, salal (*Gaultheria shallon*) at the shrub layer, lavender (*Lavandula* spp.) at the perennial layer, fuchsia (*Fuchsia magellanica*) at the annual layer (or the shrub layer if you live in a warm climate), and bearberry (*Arctostaphylos uva-ursi*) at the ground cover layer.

INTRODUCTION

IMAGINE A THRIVING FOOD GARDEN. What do you see? Tidy rows of different plants. The lettuce here, all in a row, the tomatoes over there, and between each plant bare soil kept free of weeds. Perhaps, in a large garden, you have small shrubs like currants (*Ribes* spp.) or raspberries (*Rubus* spp.) in a separate section, and beyond that, in yet another area, fruit trees planted by themselves in an orchard. Each type of plant is set off in its own little place, never to mix, and the area around each plant is empty, no weeds or other plants encroaching on its space.

Now head out to a natural area and look around. How much bare ground do you see? In all but the very driest climates, nearly every bit of ground will have plants growing on it. If a tree falling or a plant dying creates some empty space, it is almost immediately filled by a new wave of plants. And how often are those plants neatly segregated by type? Instead of the discreet sections of a traditional vegetable garden, you'll find trees large and small, shrubs, herbaceous perennials, small ground covers, and vines all growing together.

If you take a traditional food garden and stop maintaining it, that bare soil and those neat divisions between plant types will disappear rapidly as weeds move in, filling in all the gaps and naturally mixing different types of plants together. Traditional gardening often means fighting against nature taking over the space and turning it back into the thickly planted, layered space it would be without human intervention. But if you look to nature when designing your garden, mixing up different plant types and filling in all the open space, you'll have to do a lot less work. Instead of fighting nature, you'll be working with it.

That's the main concept of this book: to plant up all the layers, all the spaces, of your garden with edible plants to make a different kind of more beautiful, more ecologically sound, less labor-intensive, edible garden.

Natural plant communities are filled with many layers of plants and very little bare soil. Patterning our edible gardens after natural plant communities is a way to reduce required maintenance and increase harvests while still creating an attractive space.

Now, of course, a traditional vegetable garden produces a lot more food than most natural areas, and we do need food to live. But, as I'll show you in this book, you can grow a lot of food while still designing a landscape that follows nature's model. By choosing a diverse range of plants, from all the different layers of nature, that are well suited to your climate *and* have edible fruit, flowers, leaves, stems, roots, or seeds, you can create a version of a natural landscape that will provide you with a lot to eat.

And you won't just harvest food. Your new layered edible landscape can be beautiful, a place to relax, have fun with your family, or entertain friends. It can provide all the things that feed your soul as well as your body.

These concepts aren't new. Many ideas discussed in this book take inspiration from the practices of permaculture and "food forests." But this is a simpler, more flexible approach, guiding to you to start with the garden you have and think creatively about the edible plants you can fit into the layers of your landscape. As you fill in the spaces in your garden, you'll see it transform into a beautiful, edible paradise.

I was introduced to this approach to gardening when I was in my first year teaching at the University of British Columbia Botanical Garden. At the time, I was mostly teaching edible gardening methods that relied on raised beds and tidy grids of plants, and that was how I did most of my growing at home.

That all changed when my then-manager (who I now have the privilege of calling my friend) suggested I learn a little about the food forest style of gardening. I started by just learning the terms and techniques, but I was soon completely hooked. At home, I started breaking away from my orderly grids and started creating more complex plantings that served multiple purposes, combining different plant types and layers together.

My first big success with this growing style was the apple/creeping thyme (*Malus* spp., *Thymus* spp.) bed I created in my front garden. An apple tree is a great addition to any garden: the flowers are beautiful and of course the fruits that come later are delicious. As I laid out my garden, I found that there was so much space under and around that tree to grow other things. I filled in all the other layers: the shrubs of dwarf blueberries (*Vaccinium* spp.) yielded fruit and left colorful red twigs through the winter. Greek mountain tea

In my own garden, I incorporate layers of edible plants everywhere. It has resulted in a prolific harvest, a neighborhood-friendly landscape, and a garden that almost cares for itself.

(*Sideritis scardica*) provided silvery leaves and yellow flowers and a delicious harvest of tea. Strawberries (*Fragaria* spp.) brought flowers in the spring and still more fruit, while the creeping thymes that flowered in shades of pink-mauve spread to cover all the ground. I added quick-growing kale, mustard, and Swiss chard to fill in any gaps and provide a harvest of nutritious greens. The planting was as beautiful as you could wish from a front-yard garden and produced so much food. It looked nothing like a traditional vegetable garden, or berry patch, or orchard, because it was all of those things at once, maximizing the space's productivity.

A big part of the flexibility and adaptability of the layered garden comes from the plant choices. The traditional vegetable garden relies on just a handful of species, generally fast-growing annuals, which require lots of care, highly fertile soil, and, of course, replanting each season. But those classics—the tomatoes, lettuce, beans, and squash—barely scratch the surface of the edible plants you can grow.

Throughout this book I'll be introducing you to many other species, mostly perennials, which are tasty, easy to grow, and that you may not have eaten before. There are so many cool, edible plants that you probably already have planted in your garden without realizing they were edible. While these are not the classic supermarket species, they have a lot to recommend them. Perennial species are generally slower to establish and may take more time to start producing food, but once they are established, they require little intervention from the gardener. Choosing these species allows you to do more enjoying of the landscape and harvesting food and less weeding, planting, and fussing. And many of them are not just edible, they're also beautiful and provide habitat for pollinators and beneficial insects and birds.

I've made mistakes along the way as I have experimented with this style of gardening. From being stumped by deeply shaded garden sites, to letting weeds take over, or just being too impatient in waiting for trees and perennials to start producing food, I've done my share of learning lessons the hard way. My goal with this book is to let you learn those lessons the *easy* way, from my mistakes, rather than your own, and get your own layered edible garden off on the right foot from day one.

But the key factor in designing and creating a layered edible garden is you. Your life, your tastes, your family and friends, your hobbies. You can design a layered garden over your rural acreage, or in a cluster of containers on an apartment balcony. This style of gardening is all about making a space that works for you on every level.

This also comes out in practical, utilitarian concerns, like identifying the amount of work you want to put into maintaining your plants and the types of food you want to harvest, but also emotional and aesthetic ones. If you like entertaining in your backyard, your edible layered garden can be a beautiful backdrop to a seating area around a fire pit. If you have small children, you can design your garden with places to hide and make forts, fill it with tasty foods they'll enjoy harvesting and helping you cook, while eschewing species with dangerous spines. Beyond that, the layered garden can evolve and change as your life does: you can adapt the children's garden to become an entertaining space as children grow up and leave the nest, and you can also adjust the amount of work you need to put into the space for changing work schedules and aging joints.

There is no one correct way to make a layered garden. Rather, it is a flexible set of concepts and techniques you can use to create the perfect space for you and your life.

I've been gardening this way for years now, and I am still exploring, customizing, experimenting, and having fun seeking out different plants to fit into the open spaces in my garden. It's still very exciting, and I can't wait for you to start exploring this garden style too.

From ground covers to trees and everything in between, my home garden boasts a wide array of plant layers, creating a resilient garden that is highly productive.

1

WHY A VERTICALLY LAYERED GARDEN

No matter the size of the area you have to work with, there are so many reasons why you should turn your garden space into a layered edible garden. These range from practical matters—such as using space more effectively and reducing the amount of work your garden requires— to the intangible ways it can expand your world, help you make new friends, and build quality time with your family.

LET'S START WITH USING your garden space effectively and efficiently. Nearly every gardener wishes they had more space to garden in, but often our garden design keeps us from using what we have to its maximum potential.

The traditional home garden design has different areas designated for different uses: Ornamental and decorative plantings usually in front and in areas where you socialize, and more practical, food-producing spaces often tucked away out of sight. Separating those two spaces limits the usefulness of both, giving you less space to enjoy the beauty or to produce food.

The goals of beauty and food production are not mutually exclusive, though—you can do both at once, in the same area. In this book I'll recommend many beautiful plants that also can come inside and carry their weight in the kitchen, all while blurring the lines between the different parts of your garden. There will still be areas that lean toward the decorative while others lean toward the practical, but you can subtly change the focus in different parts of your garden while still letting every space do double duty.

In a highly visible garden in front of the house, for example, you might choose to plant a crabapple tree (*Malus* spp.) with a showstopping spring floral display and then harvest its small, tart fruit to make crabapple jelly. In a less public back garden, you might choose a large-fruited apple variety that still has pretty—if less abundant—spring flowers, but gives a bigger, more versatile harvest. The goal here is to have you think flexibly about all your garden spaces and plant options in order to create a garden that gives you lots of options and use.

I take advantage of all available space in my home garden by combining edible plants from every layer together in a pleasing way.

This garden is proof that beauty and food production can go hand-in-hand, even in a very visible space.

The other way our gardens waste space is by leaving some of the space above the ground empty by not working with all the different plant layers. What do I mean by that? Walk through your neighborhood and look at the plantings in front of people's houses. Often they have little growing except for a shade tree and a lawn—the canopy and ground cover layers. You could take any of those boring front gardens and add small trees, shrubs, herbaceous perennials, some climbing vines, and even root crops growing underground. Filling in those missing layers creates so much more interest and beauty, and each one offers a chance to add something edible to the mix. Each added layer can help you get more out of your existing garden space.

Traditional food gardens usually have missing spaces in their layers as well, with each plant type kept separated in its own area, in tidy rows or grids, all surrounded by bare soil. And, in the case of annual crops like most vegetables, that space is even more barren and unused much of the year, especially during the winter or when newly planted crops are small and haven't grown to fill the space yet.

This style of gardening makes sense for big commercial growers who want to care for their plants efficiently, using large machines. For a small home garden, however, it isn't the best—or the most attractive—use of space. Rather than keeping everything separate, you can layer together tall fruit trees, berry bushes, and then the smaller vegetables, all in the same space.

Thinking in terms of layers will help you get the most from your garden, in every sense: more plants, more food, more beauty, less space that doesn't live up to its full potential.

Meet the Layers

What are the eight layers of a layered edible garden? You'll learn more about each of these layers in chapter 4 and meet some great plants from each group. For now, let's focus on the eight plant forms that can create the layers of your new edible landscape.

The eight layers of an edible landscape mingle beautifully together.

Canopy trees

The canopy layer are the big trees, the ones reaching over 40 feet (12 meters). They make up biggest, most dominant layer of your garden.

Subcanopy trees

Subcanopy trees are a bit smaller, often growing in nature at the edge of woodlands.

Shrubs

Shrubs are small, multistemmed, woody plants that make up a useful, low-maintenance part of your layered edible garden.

Herbaceous perennials

Herbaceous perennials are plants that die back to the ground each year, then put up fresh new growth from the soil the following season.

Climbers

Climbers use other plants and structures for support. They grow up walls, arbors, trees, trellises, and so on.

Annuals

Annuals germinate, grow, set seeds, and then die, all in one year. Annuals are fast growers that fill in quickly.

Ground cover

The lowest layer contains small, low-growing plants that shield soil from heat and erosion and produce edible fruits and a habitat for native insects.

Root crops/rhizosphere

The rhizosphere layer is made up of plants that produce edible parts like tubers or bulbs underground.

THE BENEFITS OF GROWING FOOD IN LAYERS

Biodiversity

Filling in all the empty spaces and layers of your garden with eclectic, practical plants is not just good for you, the gardener. Adding less common species to your garden will radically increase the biodiversity of your local ecosystem that is your neighborhood.

We've all heard about how monarch butterfly caterpillars can only feed on the leaves of milkweeds (*Asclepias* spp.) (green milkweed pods are edible, by the way, when properly cooked, and are a great addition to your perennial layer). But that kind of very specific plant-insect relationship is not at all unique to monarchs and milkweed: It's more the norm than the exception, which means that plant diversity in your garden also supports *insect* diversity.

The few insects that rely on the ubiquitous species of plants installed by every landscaper are doing great. When you get a little more creative—by adding new plants, new species, new genera to your garden—each one of those additions supports new insects that enrich your local ecosystem. And a rich, diverse ecosystem of insects means lots of food for your local songbirds and so on up the food chain. Your layered edible garden will be feeding you, but also the whole natural world that lives around and with you.

This great spangled fritillary caterpillar is feasting on a violet (*Viola odorata*) leaf. Violet flowers are also a prized edible for humans.

This corner of a layered edible garden boasts plants from five layers. A hops vine (*Humulus lupulus*) on the climber layer; roses on the shrub layer; hostas (*Hosta* spp.) and sage (*Salvia officinalis*) on the perennial layer; kale, chard, and violas (*Viola* spp.) on the annual layer; and violets (*Viola odorata*) on the ground cover layer.

Bring the Bugs

Often we think of insects in the garden as problems to be avoided, but in fact the vast majority of insects cause no problems or are even beneficial in some way. They can be beautiful if you invest the time to look at them closely. Supporting a diverse world of insects will make your garden more beautiful, and learning about them can enrich your gardening experience.

Butterflies and Moths

Though the caterpillars of immature butterflies and moths do eat holes in plant leaves, that is a small price to pay to see their beautiful adult stages. Caterpillars are also a critically important food source for songbirds. Many caterpillars can only feed on one specific group of plants—monarch butterflies on milkweed, pipevine swallowtails on pipevines (*Aristolochia* spp.)—so adding more diversity of plants can level up your butterfly game as well.

Use the tool on the Native Plant Finder website, www.nwf.org/NativePlantFinder, to get a sense of which butterflies and moths your landscape supports. You can search by your region and discover the species you can expect to see based on your location and the plants you grow.

Bees

Honey bees are the most famous pollinating bee, but everywhere in the world there are literally thousands of other species of bees large and small that pollinate flowers. Though many people are afraid of their stings, bees are generally nonaggressive, only stinging if harassed or forced to defend their nest. When they're out visiting flowers, they'll ignore you.

Wasps

Wasps get a bad reputation because a few species can be aggressive and sting humans. But don't paint the whole group with such a broad brush—there are many more wasps visiting your garden than you realize, and most of them do not sting, are great pollinators, and like to eat aphids and other insect pests.

Many of the coolest—and tiniest—wasps are called parasitoids, which are fantastic, beneficial insects. The adults feed on nectar from your flowers, and then lay their eggs *inside* pests like aphids. The baby wasps develop inside their insect hosts, then emerge as adults to sip nectar, pollinate, and lay more eggs. Growing lots of nectar plants in the garden for the adult wasps will keep lots of them around to control insect pests before they get out of hand.

Flies

A few of the fly species that like to eat trash give this enormous and fascinating group of insects a bad name. Many flies in your garden are great pollinators, and you'll probably even mistake them for bees because they mimic a bee's yellow-striped body markings. You might argue they're even better garden inhabitants than bees, because they pollinate, they don't sting, and many supplement their diet of nectar from flowers by eating other insect pests.

Beetles

A couple species of beetles, like Japanese beetles and cucumber beetles, are plant pests, but many others are great garden inhabitants. Some have a voracious appetite for garden pests like slugs. And though they'll never rival butterflies for their brilliance, take a closer look at the glossy, colorful backs of beetles and you just might discover how beautiful they are.

Maypop vine
(*Passiflora incarnata*)
is a host plant for gulf
fritillary caterpillars.
It also produces
edible fruits for us.

Many popular culinary
herbs, including
this dill (*Anethum
graveolens*), support a
diversity of beneficial
insects, including
parasitoidal wasps.

Boosting biodiversity is one of many perks of growing an edible landscape filled with many layers of plants. From birds and bees to butterflies and beneficial insects, each plant supports a different group of animals. Imagine all the life you'll nurture in your edible garden, along with your own family!

Speaking of biodiversity, it can also make your life as a gardener easier. The first step up the food chain from the insects munching on your plants are the predator insects that munch on *them*. Providing a diverse ecosystem keeps those predator insects happy and healthy. In nature there's a healthy balance between the insects that eat your plants and all the other insects that eat them. Keep biodiversity in your garden high, supporting a healthy local ecosystem, and you'll reduce the chance that any one insect pest will become a serious problem.

Above predator insects on the food chain are your local songbirds. Your layered garden will provide them with lots of insects to eat, berries and other food sources, and ample locations to nest and shelter from predators.

Soil Health

One of the other benefits of this type of gardening is minimizing soil disturbance. Pulling out and planting annual crops each year disrupts the soil a lot, and many traditional vegetable gardens rely on annual tilling to control weeds. Growing mostly perennial crops and planting densely in layers will keep the soil covered and undisturbed.

This kind of minimal soil disturbance mimics how nature works: The soil tends to stay put, except for small disruptions by burrowing animals and the occasional extreme event like a tree falling, floods, or landslides. Following nature's model is always a first good step, and there are a couple reasons why this is great for you as a gardener.

This spiral garden contains a plethora of edible plants, all supported by rich soil filled with organic matter.

The first is the organic matter you find in natural soil. Leaves, stems, roots, and other plant parts falling to the soil slowly decompose and are mixed in by the efforts of worms and other life living and working within the soil. As the material decomposes, they become humus, an integral part of your soil (and not to be confused with the chickpea spread, hummus).

Humus does wonderful things. If you've ever dug through rich, healthy soil, you know it comes apart in little clumps. These are called soil aggregates, which are bound together by humus in the soil. The spaces between aggregates allow for water to drain and roots to easily grow deep into the soil.

Added bonus: The humus acts like a sponge, holding moisture in the soil through dry spells. So, if you're looking for that desirable "moist, well-drained soil" you've heard of, organic matter is the key. And the key to keeping organic matter in the soil? Reducing soil disturbance. Disruption in the form of digging or tilling lets more oxygen into the soil, causing the humus to decompose rapidly.

Soil disturbance can lead to problems with weeds as well. Most of the plants we call weeds are considered "ruderal" species by ecologists. These are plants that specialize in quickly colonizing disturbed habitats in the wild. After a tree blows over in a forest, ripping up the soil around it, you'll find countless ruderal species emerging and speeding into growth.

Many of these weedy species use a simple method to take advantage of any disturbed, exposed soils they encounter. They produce huge amounts of seeds that blow far and wide, surviving for decades in the soil without rotting—waiting for a time when the soil will be disturbed and they can germinate to start the cycle all over again.

And how do these seeds know the right time to germinate? It's pretty simple: light. Once they are buried by falling leaves or carried into the soil by earthworms and cut off from light, many weed seeds go completely dormant and won't grow until they are exposed to light. In nature, this happens when a tree blows over and churns up the soil: Some seeds are exposed and start growing immediately.

The same thing happens every time you till the garden, or dig a hole, or even churn up soil while weeding. Turning the soil exposes new seeds to the light, creating a whole new crop of weeds to deal with. When you trade your traditional, mostly annual vegetable garden for a mostly perennial, layered edible garden, there is no regular soil disturbance, the soil stays covered and dark, and those weed seeds stay safely dormant, underground, not causing any problems.

Remember, many common weeds are edible and quite delicious. This common purslane (*Portulaca oleracea*) is one of my favorite salad ingredients.

Soil: A Tool to Fight Climate Change

In addition to providing the right conditions for the plants in your garden to thrive, soils can be a tool to help fight climate change. When plants photosynthesize, they take carbon dioxide out of the air to build their stems, roots, and leaves. When they die and decompose, that carbon goes back into the air. But when those decomposing plants are integrated into the ground as soil organic matter, that carbon is sequestered—locked away for a long time.

One study of fields in Europe found that converting a space from traditional cropland to one that grows perennial plants could lock half a ton of carbon per hectare (2.5 acres) in the ground every year. Good gardening practices alone won't solve the problem of excess carbon in our atmosphere, but reducing soil disturbance and adding a garden full of thickly layered perennials is something concrete and real you can do to help.

Perennial plants, whether edible or ornamental, lock carbon into the soil and reduce soil disturbance.

NEXT PAGE: Layered edible gardens leave no room for weeds and are hotspots for thousands of species of beneficial insects that help gardeners manage pests, making them beautiful, low-maintenance spaces.

Lower Maintenance

All of these features, from the increase in biodiversity to the reduced soil disturbance, adds up to a lower-maintenance garden. Minimal weeds, reduced annual digging out and replanting, insect pests naturally managed by the healthy ecosystem of predators that thrive there. Less time doing garden tasks that you don't like, and more time to do the gardening that you enjoy, be that harvesting, redesigning, learning about and propagating new plants, sharing with family and friends, or just sitting and enjoying the beauty.

Reduction in Overwhelm

Thinking in terms of layers can help you avoid getting overwhelmed when it comes to choosing what you want to grow in your garden. Visit a good local nursery or browse through online catalogs and you'll discover literally thousands of options for your growing zone!

You can narrow down the selection when you plan with layers you need. Are you looking for small trees? A shrub? Or a ground cover to fill in the lowest layer gaps? Focusing on plants that provide edible elements is another way to limit your options to something manageable. Red maple or sugar maple? One can be used to make maple syrup, so that may be a better choice. Need late summer flowers in a shady spot? Garlic chives (*Allium tuberosum*) are a beautiful and delicious obvious option.

Thinking this way also helps you translate inspiration you may find in other climates or gardens around the world to your own garden space. A gardening magazine, a post on the internet, or a vacation to another climate could give you the worst case of "zone envy"—that wishing you could grow something that would never thrive in your climate. But if you garden in a cold, snowy climate and lust after the kind of garden that grows in the subtropics, there are other ways to get something like what you want (beyond moving). Look at each plant in your dream garden, see how it fits into its particular layer, then find a plant that will thrive in *your* climate that fits a similar niche.

No, you won't be harvesting mangoes (*Mangifera indica*) in your chilly northern-tier garden, but a beautiful pawpaw (*Asimina triloba*) will fit into the same layer and provide dramatic tropical visuals in the garden and unique flavors in the kitchen.

Pawpaws are attractive and delicious cold-climate tree fruits.

Mangoes are a good option for the tree layer, but only if you live in a tropical climate.

Densely planted beds mean less room for weeds and less watering. This combination of herbs and a summer squash plant ticks all the boxes for layered plantings.

Layered edible gardens are places of beauty coupled with production, but only if you take the time to properly design and lay out your space and choose the best plants for the job.

WHO IS IT FOR?

This approach to gardening is flexible enough for any and all gardeners who want to combine beauty, sustainability, and biological diversity with growing edible crops. The layered edible garden draws a lot from the principles of permaculture and concepts like food forests, but my focus is really on making this technique work for any gardener in any situation. This is not a prescriptive approach, rather a collection of tools and concepts that you can apply to any garden in any space.

And don't think you have to do it all at once. You can work through your garden space by space and season by season, exploring plants and finding what works for you. In my own gardening practice, I tend to be indecisive and move plants around all the time—often this means I learn a lot by watching how they respond to their new location. Think of layering as an invitation to explore and have fun, not a strict set of rules that tie you to one method or gardening layout. No matter if your garden is big or small—if you're a total novice or a seasoned green thumb—if you explore the layered garden approach to growing food, you'll find many benefits and ways to make it work for you and your situation.

Depending on your focus, interest, and situation in life, you may want to use this gardening style to achieve different goals. By understanding and articulating exactly what you want to get from your garden, you can find your own way as you apply these the ideas in this book to make the most of your gardening space.

If your primary goal is to have a beautiful garden, you can certainly take these principles and plant ideas to work a little food production into a showstopping garden. There are lots of plants listed in this book that have enough beauty to be well worth growing beyond their food value. And chances are there are already plants in your ornamental garden that are also edible—you just haven't realized it. With a little thoughtfulness in plant choices and layout, you can harvest a bounty of food from your space without compromising on display.

If, on the other hand, you want only to maximize food production from a space and don't care as much about your garden's attractiveness, a layered garden is also a great choice. Your plant selections will be different than someone more focused on aesthetics, but thinking in layers will allow you to produce more out of previously unused corners of your garden.

Be creative and aim for originality in your edible garden design.

Most traditional food crops require full sun and very rich, fertile soils to yield a good crop, but many of the choices in this book will thrive in shade or lean soils. There are probably sections of your garden that you think aren't suitable for food production, but look again—they may offer opportunities to yield berries and greens for you to enjoy.

This is also a gardening style if you want to be creative and build a garden space that doesn't look like everyone else's in your neighborhood. Landscapes often take on a depressing sameness, with each garden featuring the same ten plants that are cheaply available at every nursery and big-box store. It's so much fun to reach for something different than those boring, overplanted species and experiment with new and exciting plants. And you may well discover that many lesser-grown varieties and species are actually better garden plants than the familiar selections in every other garden you see in your neighbors' yards.

A layered edible garden is also definitely for people who love to cook and are interested in food. The established food system that delivers food to your local grocery store—and even your local farmers' market—is focused overwhelmingly on just a few species of plants. One estimate is that 75 percent of the world's food supply comes from just twelve species of plants and five species of animals. In this book alone, I highlight over sixty species of plants, virtually none of which are available in the produce section of your local grocery store. I'm only scratching the surface with these examples when it comes to interesting and delicious things you can grow in your own garden.

A garden where every space and layer is filled allows for limited work effort, while also limiting weeds and encouraging a healthy ecosystem that will control pests. It's a practical bonus for everyone, but in particular it makes for a garden you can enjoy throughout all the seasons of your life. There are times when we just can't devote the effort to gardening that's needed, whether because we're busy at work, we're dealing with illness or raising children, or we're facing the inevitable reality of getting older.

Creating a new garden can be a lot of work. Once it's established—assuming it's properly planned and designed—a layered garden should require minimal upkeep, less watering and fertilizer, so that the garden remains a solace throughout your life, rather than a chore and a source of stress.

The Horticultural Industrial Complex

Commercial plant production has become ever more mechanized and complex in recent years. Those pots of plants for sale at your local big-box store often started their life as cuttings taken halfway around the world. They were then shipped to two or three other locations for various stages of their lifecycle before being loaded on a truck and landing on the bench in front of you. That whole system is great at producing plants at a low cost, but often these common plants were chosen for their ability to be an efficient cog in the machine rather than for their beauty, vigor in the garden, or great flavor.

Many of the more unusual or heritage plants are harder to find because they can't be shipped internationally as unrooted cuttings without wilting, or they're too tall and take up too much space on a truck. But find a source for these plants, and you'll be thrilled you did.

I'll discuss some of these plants in this book, and hopefully it will whet your appetite to go in search of more. Yes, you will have to do a little more hunting around to find them, but that's half the fun. And I promise that discovering the small, independent nurseries and garden centers in your area that produce these plants will be a benefit in and of itself, as you connect with passionate gardeners and keep discovering beautiful and interesting things to add to your garden.

Mashua (*Tropaeolum tuberosum*; see page 189) is a close relative of the nasturtium. It is not a common plant, but it sure does provide tasty edible roots. It is the kind of plant that adds both visual interest and culinary delight.

OPPOSITE: Covering exposed soil with edible plants means fewer weeds and less maintenance, especially if they are perennials that return to the garden year after year.

PERENNIAL PLANT FOCUS

The main focus of this approach to gardening is growing primarily perennial crops instead of annuals. If you're used to ornamental gardening, this will be familiar: Herbaceous perennials, shrubs, and trees are the mainstay of many a decorative space. But this will be a big change if your gardening experience is limited to growing a typical vegetable garden with tomatoes, lettuce, and similar food crops.

Annual plants are central to both the traditional vegetable garden and commercial agriculture crops. Annuals do, of course, have a lot to recommend them. They grow fast, allowing you to begin harvesting soon after you plant them. And since they don't need to store energy to use for growing the next year, many annuals go all-out when it comes to producing edible fruit, flowers, or leaves, resulting in a big harvest.

But that rapid growth and short lifespan bring disadvantages as well. Supporting all of that quick growth means annuals need more resources. That means more fertilizer, more water, and richer soil to grow in. It also means more work for you, the gardener. Planting them every year is, of course, a chore in and of itself, which is something you can avoid with perennial crops, and the annual soil disturbance and blank space for them to grow into means more weeding.

This is not to say that annual crops don't have a place in the layered edible garden. Though the backbone of your garden will be perennial, annuals can be incredibly useful to fill in spaces between perennial plantings while they mature. Again, this takes inspiration from natural ecosystems, where most of the time perennials of various sorts make up the majority of the landscape, with a few fast-growing annuals filling in nooks and crannies throughout.

Year-round interest is a huge perk of the perennial garden. An annual vegetable garden will be empty for many months of the year. Perennials, especially shrubs and trees, are present all year, making the landscape look great in every season. Black chokeberry (*Aronia melanocarpa*), for example, has stunning fall foliage, and any fruit you don't harvest will persist through the winter. Blueberries have flowers in the spring, attractive silvery foliage in the summer, great fall color, and glowing red twigs in the winter.

Such a year-round presence also means that you can harvest food many different times of the year, rather than concentrating your harvest in just one season. This applies to the food *you* harvest to eat and also to the food all the other life in your local ecosystem is able to take from your space, because a diverse, perennial-focused garden ensures there is food and habitat for insects and other wildlife throughout the year.

Growing perennial food crops is also fun. It gives you a deeper connection to the passing seasons as the plants grow and develop over years. A quick planting of lettuce is here and then gone, while your asparagus or fruit trees will be developing, getting stronger year after year, and marking the seasons for you in predictable ways as they emerge from dormancy in the spring, push into growth and flowering, and cycle back to fall color and dormancy for the next winter.

Food-producing shrubs, such as this black chokeberry (*Aronia melanocarpa*; see page 141) are an important part of your layered edible garden.

OPPOSITE: Combining perennial and annual crops together can result in some pretty spectacular combinations. Here, oregano is in bloom, in combination with kale, fava beans, Swiss chard, and marigolds.

SCALABLE AND CUSTOMIZABLE

There are lots of ways to think about plants and garden design. Defining plants as layers is useful because it makes this style of gardening modular. You can build a layered garden even if your only outdoor space is a small patio or balcony by choosing different-sized plants and growing them in containers, making good use of the climber layer to make the most of your small area.

In fact, if you are new to gardening, I would recommend starting with a very small area first, even if your overall plan is to fill a large garden space. Pick a spot, look at what layers you have, and start filling in the missing ones. Once that is done and the plants are thriving, move on to the next area and fill in the layers there. This modular approach works no matter how large or small your garden.

INTANGIBLE BENEFITS

In addition to all the great, practical reasons to create a layered edible garden, there are so many other ways it can enrich and improve your life. My garden holds many different plants, all living together, filling, and creating spaces for each other to live. It's a constant reminder of how interconnected living things are and how our activities in the garden have an effect on systems around us. My urban garden with my choices of plants—especially my decision not to use harmful pesticides—ripple and have impacts all through my local ecosystem. In a world that can feel full of ills and problems that are too big to tackle, going out into my garden and knowing I can do things—real, if small, things—that have a positive impact on the world around me is a great lesson and comfort.

Small Spaces Can Have an Impact

Often environmental issues can feel huge, and out of our reach to really impact. We try to make the right choices, but issues like climate change and habitat destruction are bigger than any one person. But your garden—big or small—can make a huge impact on the health of local insect populations. A fascinating research study in Michigan looked at bumblebee diversity in a whole range of different sites, from nature preserves to farmland to urban landscapes to vacant lots in cities. And they found that the vacant lots had as much diversity of bumblebees as the nature preserves! Why? Because no one was spraying insecticides on those lots, and nature had filled them with a diversity of layered plants. Which means you can do the same in your own garden, creating a space with nature preserve-level diversity of native bees and other pollinators and insects.

Whether you grow in a small urban backyard or on acres in the country, layered food gardens are for you. This gardener is growing annual edibles beneath the "skirt" of an old grape vine.

I'm also always learning in my garden. Its complexity, full of so many ways to make food and beauty in my space, keeps me constantly interested. I don't just grow the same five vegetables each year; I'm constantly trying new plants, researching what they need to thrive and the ways they will fit into the ecosystem that is my garden. And that keeps me learning new, sometimes unexpected, details and methods.

For example, when I wanted to add more fruiting plants to my garden, I started researching the mechanics of pollination, especially how flowers interact with insects and other pollinators in order to make fruit. From there, I turned to plant morphology: all their different shapes, forms, and the reasons they grow the way they do.

Soon I was wondering why one plant thrives in my climate but fails for a friend in a different region. Asking that question got me interested in the art of "zone pushing," the sometimes addictive method of tweaking conditions in my garden to allow plants that usually wouldn't grow in my climate to thrive.

Learning all this about plants—how they grow, how I can use them—also led me to study up on plant propagation. This is the art of making more plants for me to enjoy in my own garden or to share with friends. It gives me a chance to celebrate what I grow and spread around plants that are a little bit different from what you typically find in standard nurseries and garden centers.

In addition to the scientific understanding of plants and how they work, my garden has introduced me to culture as it relates to plants and how they are grown, interacted with, and eaten around the world. Many of the plants I recommend in this book may be unfamiliar as food plants to you, but often there is a long history of humans' relationship with them. Some are obscure because their history has unfolded on the other side of the globe, while others have been nearly lost due to colonialism. And some, as it turns out, may be unfamiliar to you but very familiar to your neighbor down the street.

Researching recipes and ways to cook from your layered garden will introduce you to countless stories of plants and cultures. Each story brought to my garden by a certain plant has enriched my life and helped me see the world in a new way.

RIGHT: Learning how to prepare and cook unusual edibles like this Jerusalem artichoke (*Helianthus tuberosus*, see page 188) introduces you to rich flavors and cultures beyond your own.

OPPOSITE: A broad selection of flowering edible and non-edible plants translates to a greater diversity of pollinators.

My garden has also brought so many friends into my life. I once reached out to a total stranger to ask where they had acquired their unusual mashua tubers (*Tropaeolum tuberosum*; see page 189 for more on this fascinating South American plant). That conversation led to other conversations, and now I count that person as a good friend, someone who inspires me continually to seek and learn about neat exotic and practical plants to grow. Two people brought together by funny-colored tubers from the Andes. It's amazing how the world works.

Lean into the conversations you'll have in person or online around growing these plants, finding sources for them, and learning the best ways to cook them. You'll find yourself connected to a whole network of wonderful people who love plants and gardening as much as you.

Those benefits will come full circle as well. Your garden will help you make friends, and those friends will help improve your garden as they share their knowledge, experience, seeds, tubers, and plant cuttings. Just be sure to pay that forward and pass your favorite plants on to others who are interested in this style of gardening.

Gardening can also deepen connections right at home. My son has joined me in the garden since he was a baby, and he now helps me select new food varieties to grow. This help has gotten more involved each year. He reads the seed descriptions and decides which new flavors he wants to try next. He started just growing quick-and-easy radishes and has graduated to potting up and training tomatoes!

In a world where the proliferation of screens can make it feel like people—even family members sitting in the same room—are miles away, gardening is a wonderful physical, in-person activity that helps us connect and spend quality time together. You can connect with family in any kind of garden, but I would argue that a layered edible garden provides the best-quality garden time possible. The beauty, the diversity, and the timeless appeal of being able to walk outside and grab a quick snack draws in gardeners and non-gardeners of all ages. It is certainly a lot more fun for everyone involved than spending hours working down the rows of a traditional vegetable garden trying to keep all that bare soil free of weeds.

Have I convinced you to transform your landscape into a layered edible garden? Then it's time to read on! Time to get into the details of exactly how you can build the garden of your dreams and start gaining all the benefits of food, beauty, a healthy ecosystem, and new connections with friends and family.

Spending time with my son is yet another joy of growing a layered edible garden.

2

BUILDING A LAYERED GARDEN

Like any garden, the first step to making a layered garden is taking thorough stock of what you already have and putting it down on paper so you can start coming up with ideas on how to improve it. That process is called making a base plan.

A BASE PLAN IS A SIMPLE, scale drawing of your gardening space, be it a patio, a small urban yard, or country acreage. You may wonder why this is needed when you can just peek out the window and see what your garden space looks like, but making a base plan will be incredibly useful.

First, measuring everything out and then looking down at your space will give you a different perspective. You may be surprised how much space is being taken up by an old, overgrown hedge that you don't like much anyway. Or you might notice that the side yard between your house and your neighbor's is bigger than you thought and could be turned into an exciting section of the garden.

Once you have your bird's-eye view of the garden created on paper, it can also help unlock your creativity and open up a playfully exploration of different ways to create a garden. In the real world, rearranging a garden takes time and a lot of hard work. On paper, you can remove a tree or add a new path with a few strokes of an eraser or pencil. That freedom can help you come up with exciting ideas.

There are a couple ways to approach making your site plan. You can get a good, long tape measure—100 feet (30.5 m) works best for most yards—and measure the space, noting the dimensions as you go. Start by measuring your property margins, then add in the major elements: house, driveway, garage, shed, and large trees. You don't need to worry too much about precision, but the more carefully you can measure the better. Using graph paper can be an easy way to make sure everything is to scale.

In addition to making one big site plan, you might want to create smaller, more detailed plans for areas with a lot of planting. One plan can be a big overall view, while another might be just the area around the house, a large raised bed, or a densely planted border.

This vibrant space is a great example of what a layered edible garden can look like. In it, the gardener has included pole beans, artichokes, borage, calendula, squash, leeks, and so much more.

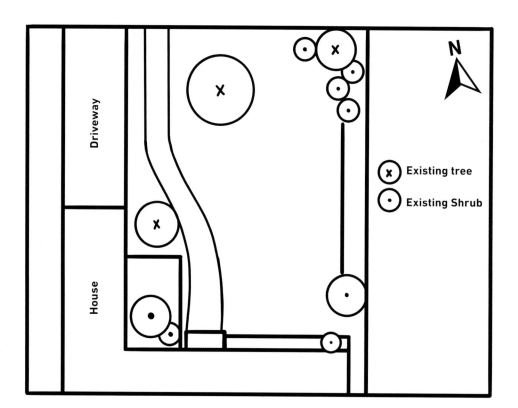

This hand-drawn base plan shows the existing plants on the property, along with features like the house, patio, walkways, and current garden beds. Make a base plan of your own property to get an overhead look at what spaces you have to fill.

Existing tree

Existing Shrub

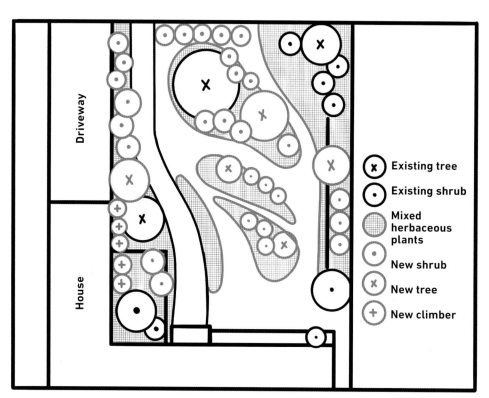

In this version of the same base plan, you can see the new plants that will be added (in red). The planting areas have been enlarged and lawn areas have been reduced.

Existing tree

Existing shrub

Mixed herbaceous plants

New shrub

New tree

New climber

Another way to make a base plan is to grab an image of your property from Google Earth or another online mapping service, print it out, and draw right on top of it to mark out areas where new plantings will go as the garden progresses.

1 Kiwis (multiple on a structure; assuming the neighbor doesn't have a tall fence casting shade onto that area)

2 Quince or jujube

3 Black chokeberry or elderberry

4 Pineapple guava or goji

5 Black mulberry or persimmon

6 Currants and gooseberries

7 Two raised beds

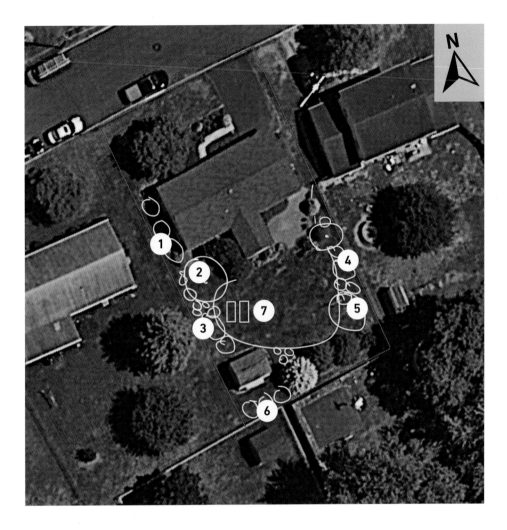

You can skip some of the measuring work by getting a map of your plot online. Most cities offer programs that give you the official measurements of your lot. Barring that, online mapping services have satellite images of nearly every location on earth. Download an image of your property, print it out, and you have a nice starting point for your site plan. You can then draw over that to mark out the precise locations of features, like trees, that might not show up on the plot map.

With your plan drawn up—or retrieved from the internet—you can start doing what is called a *site analysis*. This is taking the plan down to the next level of detail, marking smaller features like plants, structures like fences and utilities, as well as gates and entrances to your house and garage. If possible, also mark out large windows and areas where there might be traffic; this gives you an idea of what will be visible in your garden from inside and around your house. Also note where you have sun or shade, wet or dry conditions.

Here are some other elements you should consider in doing your site analysis.

Edible plants can be combined with non-edibles to block unsightly views of the neighbor's garage, shed, or pool.

SIGHTLINES

Stand in your yard and look around. Do you see something you like? A mountain or a beautiful tree in a neighbor's yard? Mark these on your plan as a reminder not to plant anything that will block the view. See something ugly, like the neighbor's trash can, a busy road, or a power line? Mark that as well, so you can remember to plant or build something to screen that view. If you have a patio or other outdoor seating area, notice the views from that spot while seated, so you can design plantings that will enhance your experience there. Also walk around and note any spots with nice views—those are good places to add a bench or patio.

The most important sightlines are also the easiest to overlook: the ones inside your home. The truth is, many of us spend a great deal of time looking at our gardens through the windows in our living rooms, home offices, or kitchens. Move around your house the way you do every day and make notes on each view you see as you go from breakfast to desk to cooking dinner to relaxing on the couch. Mark on your site plan exactly what parts of your yard can be seen from each spot. Note the window heights, too, so that you don't accidentally plant something tall enough to block the view.

It's worth taking your time with this part of the project—a beautiful garden design is of no use if it isn't situated where you can see and enjoy it!

PLANT INVENTORY

Next, make a list of all the plants that you already have on the site, from big trees to small perennials and lawn areas, marking their locations. If you're not sure what each plant is, try a plant identification app on your smartphone. The technology is generally pretty good for the most common landscaping plants but is far from perfect.

Use an app as a starting point for identifying what you have growing. Once you've finished this book and transformed your yard, you may have all sorts of unusual goodies growing that the apps won't be able to identify, but usually they do great with the standard varieties in most gardens.

TRAFFIC AND PATHS

We have to move through our landscape to destinations. Where do you walk to get to the garage? To take the dog out? Where do your kids like to play? And don't forget the other people who come to your property. Mail carriers, delivery people, meter readers, all move through your space.

Mark the current paths you and everyone else takes on your plan. The simplest solution is usually the best one: Build your intentional paths along the routes that people already take. It might be tempting to move a path in order to add a dramatic, beautiful curve or take advantage of a prime sunny spot, but the less direct a path you make, the more likely it is that people will deviate from it. If your intended path is too far out of the way, the most likely outcome is that trampling, hurrying feet will make their own paths.

Make sure your paths are wide enough to be comfortable. A path that's two to four feet (about a half-meter to a meter and a quarter) wide is a good rule of thumb. Narrower paths feel slow and intimate—good for wandering around a garden—while wider ones are better for high-traffic areas and main walkways to the house's entrances.

Don't forget the wheelbarrows and other equipment you'll need to move through and around the space. You'll want your main pathways at least three feet (one meter) wide so you can easily get in and out with the tools you need to maintain your garden.

HOW YOU USE YOUR SPACE

Where do you like to sit outside and read? Where do the children like to play and the dog run? Where do you stand when you visit with your neighbor over the fence? Do you have a spot to eat outside or entertain friends?

Mark all of these on the site plan, then spend some time thinking about the uses you'd like to make of the space. You may use a space one way right now, but that doesn't mean it's the best or only way it can be used. This is a time to get as creative as possible and dream big.

Make a list of everything you wish you could do in your space, then print out copies of your site plan and play around with all the ways they might fit into your garden.

Do you not actually use your patio? Why not? Would it be better if you moved it closer to the house or received some added shade? Would a patch of lawn for games of catch be better if it was moved farther from the house to keep the patio area more peaceful?

EXPOSURE

Moving on from the human interactions with your landscape, you should now consider the environmental factors that will have an effect on the plants you will be growing. One fundamental constraint of any gardening is the amount of sunlight that plants in each space will receive. Sun and shade are the results of two interacting factors: where you have large, shade-producing objects, be they trees or buildings, and where the sun is.

In the Northern Hemisphere, the sun will always be to the south, so the north side of any building, tree, fence, or hedge will be shady while the south side is sunny. In the Southern Hemisphere this is exactly reversed, with the north side of buildings and trees getting the most sun. This effect is most pronounced in the winter, less so in the summer, and becomes amplified the farther you are from the equator. Wherever you live, the sun rises in the east and sets in the west, so the east side of trees and structures will get morning sun and the west side sun in the afternoon.

The best way to figure out the amount of sun in your yard is to simply go out and look at it at different times of the day. Print off multiple copies of your site plan and spend a day marking where there is shade and where there is sun, a couple of times in the morning and a couple of times in the afternoon. Areas that get six or more hours of direct sun a day are considered to be in full sun, six to around three hours are areas of partial shade, and less than that is considered in full shade.

This simple formula fudges many variations, as different types of trees will cast dark or lighter shade, and the shade will change a little during the seasons, with the sun lower in the sky in the winter than in the summer, and, of course, as deciduous trees leaf out. Full sun and deep shade are always easy to identify, while the partially shady spaces in between can be a tricky, literal grey area.

Make the best measurements you can and know that you'll probably adjust those designations over time as your garden grows in and you observe your plants. And be aware that most gardens get shadier over time as trees grow and expand their canopies. Once your garden is growing, one of the maintenance tasks we'll discuss in chapter 5 involves taking stock of whether each plant needs more or less sun.

SOIL CONDITIONS

Soil is the basis of everything in the garden. Experienced gardeners think about their soil conditions endlessly. The realities of your soil composition and condition should inform your gardening practices and plant choices.

Perhaps the most important aspect of garden soils is drainage. Observe your space after a heavy rain and mark any areas where water puddles up and is slow to drain away. To get a more precise look at drainage, you can dig a hole 12 inches (30.5 cm) deep, fill it with water, and watch how long it takes to drain away. If it lingers for over twenty-four hours, you have a poorly drained site. If it disappears in a matter of hours, you have a very dry, well-drained site.

Trees and structures can modify the water availability in soils as well. Roof overhangs can create bone-dry areas next to the house and some tree species—maples famously so—have thirsty root systems that quickly absorb any available water from the soil surface. Conversely, downspouts from roof gutters can direct significantly more water to specific areas in the garden to make them wetter.

With the right mindset, these can be opportunities rather than challenges. Areas that are drier or wetter than the rest of your garden are chances to expand the diversity of plants you can grow and enjoy.

You can also modify your soils to make them stay wetter or help them drain faster. Adding organic matter such as compost and mulching will help retain moisture, and you can do a little land sculpting to help hold water. Terraces on a slope or raised berms around the edge of property will help any rain that falls stay on the property and soak into the soil rather than simply running away downhill. And, of course, you can add irrigation to artificially provide water where nature doesn't.

Don't forget to add paths to your edible garden. They make it easy to meander and to harvest.

NEXT PAGE: Choose plants that suit your site. This garden receives full sun all day, so it is not the place to plant shade-loving edibles.

When wanting to increase drainage, the temptation might be to dry and dig well-drained materials into the soil. However, even if you removed all your heavy clay soil, which retains water, and replaced it with pure sand, which is good for drainage, if the surrounding area is made up of poorly drained soil, all you'll have done is create a swimming pool full of sand. There needs to be somewhere for the water to go.

You can try two options to increase drainage. The easiest is to build up raised beds or informal raised areas using a well-drained soil mix (see page 63). The other option is to dig pipes into your garden. Known as drainage tiles, you can place them by digging trenches that are sloped away from your garden, putting in a layer of gravel and a drainage pipe, and then covering them up. These will allow excess moisture to drain away from your beds.

As in all things gardening, however, you'll save yourself from work and headaches by using the soil conditions nature gave you as much as possible. There are many beautiful, edible, plants adapted to growing in wet areas and a whole other world of those thrive in lean, droughty soils. Embracing what you have

will allow you to select plants that grow best in your natural conditions.

If you have wet soil, you can build a few raised beds to provide a spot for dry-loving plants, while also enjoying all the things that love your conditions. Added benefit: You won't have to water much in the rest of the garden. And if you have dry soil, pick a spot to irrigate and grow your water-hungry plants there, and enjoy the fact that many popular garden plants, including many fruit trees, are native to dry climates.

Next, get a soil test. You can find independent laboratories and, depending on where you live, local government offices that offer inexpensive soil testing options. Kits are also available to test the soil yourself, but it is generally best to just send your soil to a professional lab for analysis. Since the cost of a do-it-yourself kit and what a professional lab charges are about the same, the lab option is probably the best, as it will give you more accurate results along with specific recommendations based on the test results.

Soil tests give you useful information that you can't get any other way. You'll learn the amount of various nutrients in your soil and your soil's pH (a measure of soil

Low-lying areas of the landscape that are poorly draining are not the best sites for food gardens, unless you can install raised beds or swales to divert the water to a different spot in the garden.

Terraced gardens are a great way to optimize food growing on a sloped site.

acidity), and the tests usually include recommendations for improving those conditions. In general, the layered garden approach doesn't use fertilizers or other inputs, instead relying on increasing your soil fertility naturally by allowing plant material to decompose in place.

There are times, though, when you really need to know what's going on in your soil. For instance, some soils have serious nutrient deficiencies. If you get test results that indicate this, adding organic fertilizer can help compensate and ensure that your plants will thrive.

Additionally, if you're gardening close to a house that was built before the 1970s, you should test for lead. This substance was a common ingredient in house paints in the middle part of the last century, and lead from that paint may have flaked off the house into the soil. Breathing the dust from lead-contaminated soil is the main way it gets absorbed into our bodies. If your test shows lead in a section of your yard, make sure to place children's play areas away from the areas with lead contamination. Keep that soil covered with mulch and plants and disturb it as little as possible—an easy thing to do with the layered garden approach.

There are some edible plants that thrive in poorly drained sites, such as these cranberries (*Vaccinium* subg. *Oxycoccus*). If you can't modify your soil and site to drain better, perhaps opt for plants that tolerate boggy locations.

Gardeners who live in dry-climate locations will need to choose their plants accordingly. This wolfberry shrub (also known as Anderson thornbush, *Lycium andersonii*) is one such option. Seek out edible plants native to your growing region if possible, so they are better adapted to your existing soil conditions.

CLIMATE

Understanding the climate where you garden is key to growing a successful garden. A plant or even a weed that is easy in one region can be all but impossible to grow somewhere else. The more you learn about the conditions you garden in and choose plants that will thrive in those conditions, the easier your gardening life will be and the more abundant your harvests.

Many different factors contribute to climate. One way to start evaluating your climate is to observe how cold your climate gets in the winter. A hardiness zone map breaks areas down based on their average low temperatures in winter. This concept was first developed in the United States by the Department of Agriculture, and it has since been adopted in Canada; you can find similar maps for zones in Europe as well.

Hardiness zones are determined by taking the lowest temperature experienced each winter for each year over the last thirty years and averaging them. Each zone is a 10°F (12.2°C) range of average winter lows. When a garden is in zone 8, it means that, over the last thirty years, the average winter low temperature has been somewhere between 10°F and 20°F (-12.2°C and -6.7°C). Zone 7 averages temperatures 10°F (12.2°C) colder, and zone 9 averages 10°F (12°C) warmer.

Understanding your average winter lows is very useful and can help when it comes to deciding which plants will thrive unprotected in your winters. Don't assume that this is the only factor that matters in your climate, however; planting by zone is far from perfect. I garden in Vancouver, Canada, and places in Arizona, Florida, central Spain, much of England, and Japan experience similar average winter low temperatures.

All of these places have distinct climates, with different amounts of summer heat and varied amounts and patterns of rainfall. For example, where I garden in the Pacific Northwest of North America, our winters aren't that cold, but they are chilly and wet, and we often got more damage from winter root rots and frost heaves than areas with much colder, but drier, winters.

So, in addition to winter cold, you should ask other questions about your climate.

Snow, sleet, and freezing rain can damage edible trees like this persimmon (*Diospyros* species) that is still heavily laden with fruits late in the autumn.

Snow cover. If you live in a cold climate, the amount of snow you get in the winter is a huge factor. Snow is an excellent insulator, trapping air pockets that form nature's bubble wrap, holding in heat and sheltering their contents from cold winds. If you have a consistent layer of deep snow all winter, this will provide excellent protection from cold extremes for your ground cover and herbaceous perennials. It won't help everything you plant, however, and it can even damage trees and shrubs that become bent or broken by a heavy layer of snow.

Summer heat. Some plants thrive best if given a long, hot summer to grow in. Others will fade away if the weather stays too hot for too long. Often hot, humid nights are the biggest factor in stressing out plants that don't grow well in hot weather.

Moisture. Rainfall is an obvious consideration for your garden planning. Plants that thrive in a desert will be completely different from those that grow in a rainforest. But don't just look at the amount of rain you get; seasonality matters too. For example, the climates on the west coast of North America, South Africa, the Mediterranean, and parts of Australia all have rainfall strongly concentrated in the winter months, counterbalanced by very dry summers. Elsewhere, many other parts of the world receive most of their rain in the summer. Plants native to each of those areas are adapted to the rainfall patterns they have evolved and will grow best in a climate with similar wet seasons.

Microclimates. In addition to your local climate, there are also microclimates. These are smaller variations in conditions that occur through your garden space. The side of your house where the sun blazes all day will be significantly hotter than the spot in the yard that is mostly full sun but gets some cool shade in the afternoon from a nearby tree. Sunny walls, especially stone or brick ones, serve as a heat sink, absorbing heat during the day and releasing it at night as temperatures drop.

The walls of your house might make even warmer microclimates, as heat from indoors seeps out through your insulation. Observing where snow melts first and where you get your first frosts can help you get a sense of the warmest and coldest parts of your garden. If you like to quantify details like this, buying a few thermometers and moving them around the yard can give you a more precise look at what areas are warmer or cooler.

Moisture levels vary in microclimates as well: Low areas where the rain collects stay wetter, while the tops of slopes stay drier. This effect is significant even if the actual elevation change is small. So don't plant rosemary or other heat-/dry-loving Mediterranean plants at the base of a slope (even a small one), as you'll increase the chance of root rot from overly moist soils. If possible, site these types of plants at the top of a slope or the highest point in a bed to allow for adequate drainage.

Cold and moisture work together, because cold air is heavier than warm air. At night, cold air collects in low-lying areas, making chilly spots called frost pockets. This makes it doubly important to position plants that love warm, dry conditions at the tops of hills and to use low-lying areas for those from cool, wet climates.

ABOVE: Mediterranean natives, like this rosemary (*Salvia rosmarinus*), do well at the top of retaining walls, where the soil drains more freely.

OPPOSITE: This olive tree is right at home next to heat-retaining stone walls. The gardener has created a microclimate that protects the tree from cold night-time temperatures.

WIND

The movement of air is another big factor in how you and your plants will react to the local climate. On a hot summer day, a gentle breeze can be pleasant, but strong winds can make a garden uncomfortable to be in. Similarly, in the winter, cold winds can cause enormous damage to plants, especially anything evergreen. Wind generally comes from the same direction—called the prevailing wind—so when you are in your garden on a windy day, notice where the wind is coming from and consider adding something to slow that wind down. A hedgerow will reduce or mitigate wind more effectively than a solid wall, and if you plant a hedge with shrubs that produce edible fruit, you'll be able to get double duty from it—triple duty if you count the fact that birds love informal hedges to shelter in.

Remember that climates are changing as well. They're getting warmer around the world, but at this point they're also more unpredictable. Maybe your neighbor tried growing a plant a decade ago and it froze over the first winter, but you may find that it will survive your winters today. Consider adding more heat and drought tolerant plants to your landscape to adapt to what your summer climate is becoming.

Once you have an idea of your local climate in terms of heat, cold, and rainfall, you can use that information to find plants suited to your conditions. Later in this book, you'll find plant profiles and basic guidelines on the conditions a particular plant loves. When looking for information online, focus on plants native to climates similar to your own or those recommended by people gardening in similar climates.

And, of course, local information is golden. What you learn from local gardening friends, garden tours, area botanical gardens, or knowledgeable garden centers and nurseries can help guide your plant selection. If you see a plant thriving in your local climate, forget anything you've read about it in books or online. Climate and a plant's ability to adapt is more complex than any one book or gardening expert can capture.

Sometimes, you just have to experiment! Just because no one around you is growing a particular plant doesn't mean it won't thrive in your garden—maybe no one had the idea to try it. And even if your gardening friend up the street has tried and failed with a plant, there are a number of reasons they got that result. Maybe the plant isn't suitable for your climate—or maybe they happened to buy an unhealthy plant, forgot to water it, or didn't plant it in a suitable location. Half the fun of gardening is trying things and seeing what happens, so don't be afraid to give a plant a try and hope for the best.

The fruits of an edible hedgerow shown here include rosehips, filberts, elderberries, blackberries, chestnuts, apples, and hawthorn.

NEED AND WANTS

At this point, you should have a good idea of what's in and around your garden, the buildings and trees that surround it, as well as your climate and exposure. Now comes the fun part: figuring out what you want from your garden space.

List everything you dream of for your space. Think big here. Yes, you probably won't be able to fit everything you want into the landscape, but start listing everything you *could* want.

Do you want lots of fresh fruits and berries? Fresh greens and other vegetables? Herbs? A dedicated pollinator bed or even a meadow? Don't forget the other factors as well: places to entertain friends, a spot to meditate, wind chimes, bird feeders, hammocks—whatever you can dream up, put it down on your list. Be sure to collect feedback from everyone in your household and learn what they would like in the garden.

As I said, you might not be able to add everything on your wish list, but giving yourself permission to dream big can open yourself up to new ideas.

WORKING WITH WHAT YOU HAVE

Once you've dreamed big, it's time to come down to earth and figure out what is possible, and what constraints you need to work around. Large structures are difficult and expensive to move, and mature trees are beautiful and take many decades to replace. When possible, respect what is already there and keep it in place, though do consider what would happen if you tore down the old shed you don't use.

It's helpful to balance the usefulness of what is present in your space—along with how you can save money and work by planning around those items—with new ideas and configurations that will freshen up your environment and your perspective. You may find that, if it really would improve the garden, making those big, expensive changes might be worth it.

IDENTIFYING CONSTRAINTS

In addition to the large features of your garden that you want to keep, there are many things that could constrain what you can do where. Trees that will grow large can't be safely planted under power lines. If you are in a rural area, deep-rooted plants can't be grown over the septic leech field. And, perhaps most importantly, take a good, hard, realistic look at your available budget of both time and money. How many hours and how many weekends are you really going to want to spend building a new patio or digging out an overgrown shrub? Remember that everything always costs more and takes more time than you first estimate, so build plenty of margins for error into your plan and what you want to try.

One approach is to break your dream project down into a series of smaller parts. Look at your plan, prioritize your projects, and take it step by step. Perhaps your first task will be planting serviceberries, with beautiful flowers and delicious fruit, to shade your sitting area that you don't use currently because it's too hot and sunny. Do this first and save replacing the inedible boxwood hedges around your house with tea shrubs for next year.

Breaking down your overall master plan into a series of small, attainable goals also allows you to have more fun along the way, enjoying and learning from each part as you finish it, then moving on to the next. This will save you the disheartening, exhausting job of working on the whole space at once, dealing with the chaos of half-finished construction for what may wind up being years.

Work out your hardscape first. Working on these areas, such as patios and paths, can damage the plants around them. After hardscaping, prioritize the larger investment plants that help anchor the space and plantings. Fruit trees and larger specimens tend to grow slower and be more expensive, so if you know you want

Greek mountain tea (*Sideritis scardica*) is a favorite experimental crop of mine. Don't be afraid to grow new-to-you things.

a particular large fruiting tree, get it in the ground as soon as possible to establish before moving on to the faster-maturing layers.

Getting these big-ticket items done first helps you visualize the space, making the projects seem less daunting. It can also help you manage your budget. You don't want to overspend on fast-growing and easy-to-acquire herbaceous plants and then run out of funds for the longer-lived plants you will potentially enjoy for decades.

Be aware that the plan for your space will probably change over time. After you have a couple seasons to actually experience your garden, you'll see where you want other improvements, like paths or seating, and have a better sense of which plants you want to add for their aesthetic and edible contributions. Your life and taste will change with time as well. Don't be afraid to lean into that! A garden is a living, dynamic space, and should work with you in every stage of your life.

In urban areas, there are other constraints you need to take into account. If you live in a neighborhood with a homeowners' association, there may be strict rules about what sorts of plantings are allowed, especially in the front garden. Most cities will also have rules regarding fence heights and locations, and often general requirements about the upkeep of front yards. Looking up those rules *before* you begin and building your plan around them will save you headaches and future conflicts.

Water is a final constraint you should consider. Depending on where in the country you live, water may be cheap and easily available, or expensive and unsustainable. Planning a water-hungry garden in areas that are prone to drought and water bans is going to set you up for stress and disappointment down the line. Also, consider where you have access to water on the property. If the only hose hookup is right by the house, and you put water-hungry plants down at the opposite end of a property, you are going to be cursing yourself every time you have to haul either hoses or water cans out there to keep those plants happy. If adding another hose connection isn't an option, plan to keep water-hungry plants close to water sources and more drought-tolerant ones farther away.

Edible plants can replace common landscape plants, often without anyone even noticing. This hedge of tea plants (*Camellia sinensis*; see page 129) is a great replacement for non-edible boxwoods if you live in a climate that's appropriate for their growth.

Large edible plants, like this hazelnut tree (*Corylus avellana*), should be incorporated into your garden's design before moving onto smaller plants.

OPPOSITE: This drought-tolerant edible garden contains species that are suited to low-water areas. It is also well mulched to preserve water within the soil and contains many perennial crops that have deeper root systems than many annual options.

SITE PREPARATION

With your plan, all your dreams, constraints, and steps in place, it's time to break ground and get space ready to add your plants. There are two main approaches you can take: making in-ground beds or building raised beds. Both approaches have upsides and they aren't mutually exclusive. You can mix and match both approaches to get the best out of different parts of your garden.

In-Ground Beds

Growing plants in regular in-ground beds has many perks, mostly because this is the simplest and cheapest solution. In-ground beds are probably the right option for you if you have a large area you are planting, if your native soil is of good quality and drains well, and if you want a more informal look for your landscape.

You could just grab a shovel and some plants and start popping them into the ground. But please don't! Planting without carefully preparing the site sets you up to spend the rest of your gardening time fighting weeds. Taking the time to properly clear the site of existing vegetation will ensure you have a garden to enjoy rather than one you spend hours working in.

The most important part of soil prep is removing the existing plants growing on the site. A weed is simply a plant growing where you don't want it, so even though turf grass is not a weed in a lawn, it quickly becomes one when it grows into your garden beds.

Before you plant a garden, clear out those plants and make sure they don't become a problem. Once you've planted it, however, you'll have to spend hours carefully pulling each individual weed, so thorough bed prep can be a huge timesaver. There are a few ways you can go about clearing existing plants from a bed, each with advantages and disadvantages. Choose whichever works best for you, but don't skip this step.

Tillage

This is the traditional, old-school way of clearing ground. You get a tiller, or even a hand shovel, and churn up the soil.

Physically chopping up and burying weeds will kill most of them, but this method has some disadvantages. First, not all weeds can be killed this way. If you till some plants with tough, spreading rhizomes, like bindweed or horsetail, you'll just propagate and spread them through the soil. That's a lesson I learned the hard way! So identify what you have growing before you till.

Tillage also damages the soil structure, makes soil prone to erosion by wind and water, exposes buried weed seeds to light (thus triggering them to germinate), disrupts the network of soil life from earthworms to beneficial fungi and bacteria, and adds excess oxygen to the soil that causes organic matter to break down more quickly.

Numerous scientific studies have shown that long-term, annual tillage is terrible for the soil. That said, using this as a one-time method to clear ground for planting and then following up with a planting of perennial crops, constant soil cover, and minimal soil disturbance can minimize negative impacts.

Tilling is one of the fastest ways to create a new planting bed for your layered edible garden.

Tilling a bed is hard, physical labor. On the plus side, it's the fastest way to prep a bed. You can till a garden bed, mulch, and plant it all in one day. You'll probably be exhausted at the end of it, but if you have lots of energy and little time, it can be a good option.

Deep Mulching

This method is simple, and it's the best choice for maintaining your soil's health. Mow the grass or other plants in your new garden area as low as possible to the ground, then pile a *thick* layer of wood chip mulch over it, 12 inches (30 cm) deep. Wood chip mulch allows air and water to pass through, preserving soil life and, as it decomposes, adding valuable organic matter to the soil.

There are some downsides, however. First, it takes a lot of wood chips. If you live in an urban area with many trees, you may be able to get these for free from arborists removing trees or clearing storm damage. Tree services produce wood chips by the ton and need somewhere to dump them. You can make friends with a local operation or use the ChipDrop website (getchipdrop.com) to arrange for free delivery of wood chips. If you're outside an area where tree companies operate, you may have to pay for wood chip mulch—and that gets expensive fast, especially if you have a large area to plant. And don't forget that you have to shovel all those wood chips into place.

Another concern is that you may not be able to smother all unwanted vegetation. While average lawn grass will never be seen or heard from again, some rhizomatous grasses, like Bermuda grass, will push

A deep layer of wood chips can be used to smother the existing vegetation and create a new planting bed. Then, new edibles can be planted right through it.

through as much as 2 feet (60 cm) of wood chips, as will vigorous perennial weeds like burdock, bindweed, or Japanese knotweed. (Burdock and Japanese knotweed are edible, in fact, but that may not be enough to change your opinion of them as weeds.) In general, deep mulching is not for the impatient gardener, as you need to ensure the weeds underneath are completely dead before you can plant.

Finally, when it comes time to plant, you need to clear away some of the wood chips to reach the soil to dig out and plant. If you're planting things like trees and shrubs, which are spaced widely, this is not a problem. If, however, you want to plant a dense border of perennials or annual crops, deep mulching isn't a practical solution.

It's great for planting an orchard or shrub border, but where you are going to put in smaller, more closely planted layers of your garden, you'll need to choose another method or wait a few years for the wood chips to break down enough to be able to plant into them more easily.

Sheet Mulching

This is similar to deep mulching, but with a twist. Instead of just piling wood chips to the sky, you first put down a layer of cardboard or newspaper (usually cardboard, which is easily available in today's world of online shopping), then cover that with a layer of wood chips or other mulch an inch or two (2.5 to 5 cm) thick, just enough to completely cover the cardboard. The cardboard layer blocks all light and thus kills everything beneath, even the vigorous plants that can push through a deep wood chip layer. The mulch layer keeps the cardboard in place and retains moisture, allowing the cardboard to quickly break down.

The biggest advantage to this method is that it's easy. No digging or hauling tons and tons of mulch. Simply collect some old boxes and get enough mulch to cover them.

One disadvantage of this method is that the cardboard layer stops the free movement of water and air to the soil below, which harms the life in the soil. It isn't as disruptive as tillage, but it's still not great. But, as with tillage, the impact is not usually serious as a one-time event. The cardboard will break down completely within the first year; after that, the soil life will quickly recover.

The other disadvantage is time. You can use this method to plant trees and shrubs immediately, putting the plants in the ground first, then sheet mulching around them to kill the surrounding weeds. But for the smaller layers of your landscape—the perennials and annuals—you have to wait for the cardboard to break down before you can plant.

To use this method to make a fully layered garden, I recommend installing the trees and shrubs, then sheet mulching, then waiting at least six months before planting the other layers of the landscape.

Smother with Plastic

For this method of vegetation elimination, take a sheet of opaque plastic, such as a tarp, and spread it over the bed-to-be. Leave the covering until all the plants under it have died. This usually takes a few weeks, depending on the temperature and time of year. Make sure everything is fully brown before you remove the plastic. When you remove the plastic, spread mulch, and plant away.

You can also use this method to reduce the number of weed seeds in the soil. Instead of mulching and planting immediately after removing the plastic, leave the soil uncovered for a week or two so that all the weed seeds at the surface have germinated. Then pull the plastic over it again to kill them. This is great for areas where you do want to grow annual crops, because it allows you to clear out many of the weed seeds that might be a problem for you in the future.

This is another very low-work method that quickly preps the soil for planting. There are downsides. For one thing, plastic isn't environmentally sustainable. And, like sheet mulching and tillage, it does do short-term damage to soil life by cutting off normal air and water flow.

Solarizing

This is similar to the above method, but instead of using opaque plastic, you use clear plastic to trap the heat of the soil like a greenhouse, killing the weeds below with heat. The end result is basically the same, but it's faster and does a little more damage to the soil life, cooking them along with the plants you want to eliminate.

This method is very effective in hot, sunny climates, less so in the cool, cloudy parts of the world.

Herbicide

Use a chemical herbicide to kill existing weeds. The most common chemical would be glyphosate or its organic alternative, horticultural vinegar. Both will kill existing weeds, though there are some differences: Glyphosate is more effective at killing all long-lived perennial weeds, while horticultural vinegar burns off the top growth without having a lasting impact on stronger perennial weeds.

These products are the least disturbing to soil life, but the obvious negatives are the safety of using chemicals in a garden space. The relative safety of glyphosate has been debated widely, and horticultural vinegar, though organic, is still a strong acid that can burn your skin and needs to be handled with care.

My recommendation is to skip them both, since there are many other effective options to clear the existing plants from your garden bed that don't involve compromising safety.

Solarizing soil by covering it with a layer of clear plastic kills the weeds beneath it, though it is tough on the soil life in the top few inches (cm) of the soil.

Raised Beds

A raised bed is just a frame of some kind filled with soil. Often built of wood, raised beds can be purchased, or you can make your own, using metal, brick, stone, or just about anything else that holds its structure. Remember that untreated wood frames will rot and need to be replaced after several years. Treated wood or naturally rot-resistant woods like cedar will last longer. Metal, stone, or brick beds should last forever, or at least as long as you'll need them.

Wood is usually a safe bet for raised beds, since modern treated lumber doesn't include chemicals that will leach into your soil, so you can use it for your edible beds without concern. You can also make more informal raised beds just by mounding soil up with no frame at all.

The height of a raised bed is really up to you, depending on what you want to get from it. One just 4 inches (10 cm) tall will give you many desirable benefits. If you want beds that are easy for someone with limited mobility to work in, or you want space for deeply rooted plants, you'll want to make your beds taller. The height is really up to what you need and want, but remember that taller beds are more expensive, as they require more materials to build and more soil to fill.

Building raised beds has a lot of advantages, but there are also a few downsides. First, they can be expensive. You may be able to mitigate this by salvaging materials you already have, but you'll need something to build the frame of the raised bed, in addition to the soil to fill them. They're also a bit more work, because you have to build the frame and haul wheelbarrows full of soil to fill them up.

But raised beds are wildly popular—and I love mine—for many reasons. Below are the benefits this option provides.

Drainage

If you want to garden in a wet space where the soil stays soggy or water pools up after a heavy rain, raised beds are the best—and sometimes only—solution. Fill your raised bed with a well-drained soil mix and the excess water will easily drain away.

As an added bonus, plant roots can grow through the well-drained soil of your raised bed to access water in the heavy soil below without the risk of being flooded out.

Weed Control

Your existing garden soil is full of weeds in two forms: the seeds of fast-growing annual weeds, and the roots and rhizomes of perennial weeds. Building a raised bed allows you to literally bury those problems and start with fresh soil that should be completely or nearly weed free.

For added insurance against perennials pushing up through the new bed, throw down a layer of cardboard at the bottom of the new raised bed before adding soil. Wind, birds, and sometimes poor gardening choices could introduce weeds and weed seeds in the future—for example, sometimes a plant you thought was great turns out to be a little too vigorous and becomes a weed. As long as you stay on top of your weeds before they get big enough to produce seeds or send rhizomes out far and wide, your weeding will be minimal.

Raised beds have many benefits, though they also require more work and more money to set up. This one contains a berry garden.

Ease of Work

If you hate getting down on your knees to garden, a raised bed makes sense. Very tall raised beds can be perfect for older gardeners who don't want to bend over or have limited mobility. Gardeners in wheelchairs or with other conditions that make it difficult to get down to ground level can enjoy gardening in raised beds with relative ease. Even if you're a young gardener now, remember that age catches up with gardeners, too, so building taller beds that are easier to work in can be great as you plan for the future.

Pets, Children—and Some Pests

A raised bed makes it much less likely for your plants to be trampled by the pets, children, or other garden visitors who may not be paying close attention to where they walk. A ground-level bed is easy to step into, but a little raised border makes a huge difference in encouraging foot traffic to move *around* the bed rather than *through* it.

How high to build your bed depends on the animals and people in your life and garden. Other animal residents in a garden can be discouraged by tall raised beds as well. Most rabbits won't jump into a bed that is more than about 1 foot (30 cm) tall, and groundhogs are discouraged by ones taller than about 3 feet (90 cm). Throw a sheet of wire mesh like chicken wire or hardware cloth on the bottom of the bed before you fill it and you can exclude burrowing mammals like voles and gophers.

Aesthetics

Finally, raised beds can look really nice. Putting your garden in raised beds gives everything a tidy, intentional look, and the materials you choose to build them can be attractive as well. Particularly in the winter months, when the garden may be a little more empty, raised beds give a nice structure to your landscape.

When it comes to aesthetics, don't be afraid to get creative with shapes. Raised beds don't have to be set up as a series of rectangles in a row. You can make interesting shapes and arrange them in patterns around your garden paths, making them as much garden sculpture as they are practical growing spaces.

If you want to soften the look of the edges, consider planting low-growing herbs like thyme as foundation plants around parts of the outer perimeter. Depending on the height of the raised bed, trailing plants like garden nasturtiums tumbling over the edge add lushness and edible flowers to beautify the space.

OPPOSITE: A gooseberry bush pruned into a standard is skirted by kale and chive plants to create a petite yet productive garden.

NEXT PAGE: I have incorporated several raised beds into my edible yard. I tend to grow primarily edible annuals and biennials in them.

Layered Gardens in Containers

Every garden needs some containers. A group of containers may be the entirety of your garden if you live in a condo or apartment with a patio or balcony. Or they can be a perfect complement to your in-ground and raised beds in a larger space.

Have more driveway than you're using to park cars? Put a bunch of containers on it and make an instant garden. Want to grow something that won't thrive because your soil is too acidic, too alkaline, too wet, or too something else? Fill a container with that plant's favorite soil mix and you're ready to go. Want to add quick interest to a boring garden space? Containers placed out in beds can add height and color in an instant.

You can follow the layered garden model in containers, though you won't be able to grow all the layers in containers. Canopy and subcanopy trees are simply too large for containers, but shrubs will grow well along with other, smaller layers.

You can achieve the mix of layers in a container garden two different ways. You can plant different layers together in the same pot, with shrubs, herbaceous perennials, and a ground cover draping over the edge all growing together. Another, and sometimes more flexible, option is to put each plant in its own container and then group them together in the same space. This choice is nice because you can rearrange plants during the growing season and easily play with different combinations to see what works aesthetically and practically.

Half whiskey barrels make great containers for edible crops.

This collection of containers sits by a front door. In it, you'll find a fig tree, various herbs, lettuce, pole beans, and more.

Containers are a great space to try ideas out. If they don't work, it's easy to rearrange or replant. And if they do, you can then take what you have and translate it into a more permanent planting in the ground.

Here are a few things to keep in mind when container gardening.

- Size matters, and bigger is better. Larger containers allow plants to develop large, healthy root systems, buffer plant roots from temperature and moisture fluctuations, and can help anchor the plant in case of heavy winds. Small pots will stunt plants and require constant watering during hot summer weather.
- Fertilizer matters too. Your plants in the ground or raised beds can send roots far and wide to get nutrients and water. Your container plants can't, so in addition to regular watering they'll also need regular fertilizer.
- At the same time, don't be afraid to get creative—anything that can hold soil can be a gardening container. Traditional pots will work, but if you like hitting thrift stores and repurposing vintage items, this is your place to let your creativity fly free.
- Remember that every container needs to hold soil—the more the better—and have holes to let water drain out. Also, not all nontraditional containers will last a long time, as wood will rot and some metals will rust and possibly leach with prolonged exposure to moisture.

(continued on next page)

(continued from previous page)

Winter Container Care

Container gardens need special care in the winter if you live somewhere with hard freezes. The containers themselves might not withstand winter weather. Porous ceramic containers, like those made of terra cottas, will absorb water; when that water freezes, it expands, shattering your pot. Most garden centers will have pots marked as "freeze safe" that can be left outside all year. If you aren't using freeze-safe pots, make sure what you have is kept either warm or dry during the cold months. Once the freezing weather arrives, move them into a garage or basement to keep them out of the cold and wet.

When growing perennials in containers that can take freezing weather, you need to take some extra precautions to protect them from winter cold. Normally, plant roots are underground, and the soil provides excellent insulation, protecting them from sudden temperature swings and the most intense cold. In a pot, above ground, roots get no such protection.

One option is to move the containers into a protected location. An unheated garage or shed is perfect, as these spaces provide insulation from the most extreme temperatures but are still cold enough for the plants to stay dormant instead of attempting to grow through the winter. This can be a perfect spot for perennials that you want to grow but aren't quite cold hardy for your climate. You can also leave pots outside, but try to give them as much insulation as possible.

Cluster all of your pots together and move them up against the side of a building. A warm, south-facing wall sheltered from excess moisture is ideal. Packing fall leaves around the outside of the pots will give some extra insulation, and snow, if you get it, is a wonderful insulator in its own right.

Other insulation methods can also be helpful. Many small coffee roasters will happily give away their empty burlap sacks to gardeners. Ask nicely and you'll be able to grab some to wrap around pots and the base of your perennials for added protection.

If your pots are just too big and heavy to move to a new, more protected location for the winter, leave them in place and either grow annuals in them or choose perennials that can survive more extreme winter temperatures.

Laying out and building your garden may not be the most enjoyable part of this process—we're always in a hurry to get to the fun, delicious plants—but I hope you'll take the time to carefully go through the steps outlined in this chapter.

Skimping on the soil prep and building part steps were mistakes I made early in my time making gardens. If you haven't prepared your space adequately, you're in for an endless, frustrating battle with weeds when you'd rather be harvesting food and enjoying your beautiful garden. Spend time on getting your garden's layout right and prepare each bed thoroughly, and you'll be grateful for the work you put in for years to come.

This large container holds a dwarf peach tree, basil plants, and a few flowers. It's surrounded by other edible plants, including sorrel and even a banana.

3

PLANTING COMPOSITION

Your beds are prepared. You understand what your local conditions are and what you need and want from your space. It's time to start your planting design!

NATURAL PLANT COMMUNITIES are a great place to find inspiration for your planting design and layout. Looking at wild landscapes will show you the natural inspiration for the layered garden and make some basic points of planting design clear.

Gardens are not natural landscapes, of course; they're cultivated places that we create to meet our needs and desires, so natural landscapes are only a starting point—but a critically important one. The more we can align our cultivation practices and goals with what a site wants to do naturally, the less work and more successful our garden will be.

Start looking at nature. Then you can figure out which changes to make in order to meet all the needs and desires you outlined when you made your plan.

One key inspiration from natural communities is the way different plants fit together. Ecologists talk about niches, the metaphorical and literal spaces where organisms find a place to live in the world. We often say that plants fit into an existing niche. This is what we explored in the last chapter, looking at the conditions you have in your garden space.

In ecology we also say that plants don't just fill niches, they also create them. Start with any flat, open field, and the conditions in it will be basically the same. But as soon as you plant a tree in that space, it creates new niches in the shaded and partially shaded spaces around it.

In a woodland, large trees dominate, but other plants fill in the niches around them, some adapted to thrive in the filtered light that makes it through their branches, while others specialize in quickly growing and taking advantage of openings in the canopy created when a tree falls. Still others find their space in time, coming up and blooming early in the spring before the tree canopy fills out, then going dormant for the summer.

The layers described in this book are essentially shorthand descriptions of different ecological niches in your garden.

This large edible garden has a mixture of flowering plants and edible crops, such as potatoes, asparagus, rhubarb, squash, and pole beans, all backed by fruit trees.

This edible garden is a perfect accompaniment to the wide open view and ranch-style home. Try to create a garden that matches the style of your home and the overall aesthetics of your region.

Observing the way every niche is filled in a natural habitat is a great clue for the kinds of niches in your own garden and how they can be filled with beautiful and delicious plants. Imitating nature's design will help you utilize every inch—and month—of your garden. It will also dramatically reduce your work as well: If you don't fill all the niches of your garden with the plants you *do* want, nature will do it for you, filling the spaces with plants that you *don't* want.

In addition to natural plant communities, manmade ones can be a great inspiration. A walk through your neighborhood can be a wonderful source for ideas. Take a close look at the landscapes you pass and see what seems to be working, what isn't, and what you like and don't like.

Go a step further by visiting local public gardens and parks. If your community has them, go on garden tours, garden walks, and open garden days. And don't forget botanical gardens, as many of their collections are curated to reflect how plants are (or can be) grown harmoniously in their native/natural geographical areas: alpine regions, coastal Mediterranean, North American deciduous forests, meadows, or other, specific landscapes.

As you visit these spaces, you'll discover individual edible plants that you love and want to include in your garden, as well as combinations that look and grow well together. Your smartphone camera is a great way to record plants and plant combinations that you love. Even if your plant identification app can't figure out what plant you saw, a good nursery or garden center staff person will usually be able to recognize the plant and show you how to purchase it.

Once you have your inspiration, natural and manmade, it's time to sit and think about how this can translate to your garden space. Not every type of garden design will be right for every space. Almost all gardens are built around a building, be it a freestanding house or a condo or an apartment, and that building does a lot to set the tone for the garden design. A cute little cottage will look perfect surrounded by an informal, wild-looking garden, while a traditional brick Colonial might look better surrounded by a formal garden with geometric beds.

START WITH THE TALLEST LAYER

Because each layer you add will affect all the others, it's best to start your design with the canopy layer (trees) and work your way down through the smaller layers.

The biggest layers—canopy trees and the smaller understory trees—will have huge impacts on the whole feel and functionality of your garden. Take your time when selecting and deciding where to plant them. One of the biggest challenges is imagining how big a tree will eventually get. When you plant a little hickory nut sapling, it's hard to imagine how it will fit into your space when it is thirty or more feet tall. And it's even harder to envision how such a tree will change over time.

I recommend going out into your garden and place sticks in the ground or lay out a garden hose to mark the eventual outline of a new tree's outermost canopy edge. Look at the canopy's width of 20 or 30 feet (6 or 9 m) and see how far its shade will extend when the tree is fully grown.

It can be even harder to imagine the height of a tree in the garden, but you can try. You could erect a tall pole in the garden, but it might be easier to take a 10-foot (3-m) board or bamboo pole and set it upright where the tree will go. Standing back, you can mentally multiply the height of that pole by two or three or four to get an idea of how high the tree will grow. Will it block views? Cover up windows? Grow into power lines? It's worth taking the time to try different locations, move the pole around, and even leave it there for a few days.

Putting in a canopy tree or even a smaller understory tree is a big investment, and you want to make sure you get the location right.

Large canopy trees can and should be used as a support for plants in the climber layer. Here, garden peas are supported by the trunk of a tree.

NEXT PAGE: Fill in your garden as much as possible to keep undesirable plants from moving in. Utilize every opening to create a diverse space that takes advantage of all usable areas.

MOVE ON TO THE LOWER LAYERS

Once you have your canopy and understory trees positioned, you can move on to designing the locations of your other layers, fitting in shrubs, herbaceous perennials, ground covers, and underground crops into the spaces around your trees.

Remember that a garden design is a living thing. Don't imagine that you'll design one planting composition and call it done. You'll be rearranging, replanting, and reimagining the space throughout your time as a gardener—but that's part of the fun!

PLAN FOR CHANGE

The biggest change to plan for in the garden is the way your plants will grow. You plant small things and they get bigger with time. Hardly shocking, but it can be easy to forget. Planning for these changes will make your garden life easier and more successful.

New plantings, with room for everything to mature, leave a lot of bare soil to start, which means prime conditions for weeds. As they mature, plants can start crowding and shading each other out. A little planning and good design can help avoid either problem.

The canopy and subcanopy layers will get bigger each year, casting more and more shade, but don't get too caught up in the listed "mature sizes" for trees—trees will keep growing throughout their lives. Their growth rate slows as they age, but, no matter the size of your tree, it will keep getting bigger.

Pruning is one option to keep shade from creeping over your garden. Practically speaking, it is difficult to prune a large canopy tree, and done incorrectly it can destroy the natural form and beauty of its branching structures. Smaller trees and shrubs are much easier to handle, but don't underestimate how much work goes into annual trimming. It's easy for a pruned shrub to get away from you if you have a few busy seasons and can't keep on top of it.

The best option may be to plan your design to work with that slowly increasingly shade. One approach can be to put short-lived plants at the edge of the sun/shade zone. As they reach the end of their life, it will be time to replace them with something that loves more shade anyway.

Raspberries are a great option here, as the individual plants simply don't live that long. A long-lived plant, like asparagus, would be best sited where the shade of a tree will never reach it, as you will likely be harvesting for decades to come. Similarly, site your very favorite, most expensive plants in areas that will stay sunny or shaded (depending on what the plant requires) and choose options you won't mind discarding in the transition zone, so you can replace them without regret. Most members of the herbaceous perennials and ground cover layer are also easy to dig up and move to a new location, unlike a mature shrub or small tree, which, if the site becomes unsuitable, will simply have to be removed.

Shrubs, herbaceous perennials, and ground covers will take a few years to fill out to their mature size. One of the biggest mistakes I made early in my journey as a layered gardener was forgetting this fact and leaving the ground around those new plantings bare. All that bare space is begging to be filled with weeds! That, combined with the fact that planting disturbs the soil, stirring up dormant weed seeds, can make new plantings a weeding nightmare.

Your first defense is a thick layer of mulch and budgeting your time for the first few seasons so you can get out and weed regularly. Small, newly germinated weeds are easily pulled out. One that has gotten firmly established can be extremely difficult to remove.

You can also fill in that space with short-term plants. Not only does this give the weeds fewer places to grow, it fills up space that would otherwise be unused. Annuals are your best option here: many of the traditional vegetables, like cabbages and tomatoes, or even flowering edible annuals like nasturtiums. These will fill in the space quickly, then vanish at the end of the season. This is what we see in nature: Bare ground is filled with fast-growing, short-lived annuals and perennials, then these are slowly replaced by slower-growing, longer-lived species.

A dense planting of currants (*Ribes* species) sits at the back of an urban backyard. It is fronted by perennial edibles, including rhubarb.

The edible oca (*Oxalis tuberosa*) plants at my feet serve an important role in both the ground cover and the root crop layers, acting as both a living mulch and an edible tuber.

With a little planning, you can choose to fill empty space with fast-growing vegetables rather than pesky weeds. Use caution when filling temporary spaces with annuals that get very large, like tomatoes, as they can easily grow over and shade out the immature plants around them. Choose dwarf varieties, keep them staked, and cut back any wandering branches that start covering the plants around them. When planting close to perennials, avoid varieties with deep root systems that will compete with the roots of the perennials for resources. Often greens like lettuce or spinach work well, as do dwarf peas.

As you work through your layers, choosing the plants you want to include in your garden, it's important to take stock of what each of them will bring to your garden. Look carefully to see if there's anything you're missing.

One great way to see where you have missing parts is to take the list of plants that are currently in your garden, along with those you're planning to add, and list them on a sheet of paper or spreadsheet. Mark down the characteristics of each one. Use the following as your guide.

Features to Consider for Each Plant

Which Layer Is It in?

This is the most important question, but take stock of what you have in each layer of your garden space. And don't forget what's in your neighbor's garden space as well. Plants don't respect property lines, so trees—especially large trees—next door will still be impacting your garden with their shade and root systems.

In general, you will have more diversity in the smaller layers—the herbaceous perennial and ground cover layers especially—and less in the larger layers like canopy trees.

Often the shrub and ground cover layers are overlooked in designs, as we focus on the many beautiful and productive herbaceous perennials that we can grow.

There are so many shrubs that produce lots of delicious fruit, and these probably require the least work to maintain of any of the garden layers.

Ground covers are key weed controllers and serve as living mulch that protects the soil through the seasons from erosion, frost heaving, and drying out. These can be tucked into so many spaces to increase what you can harvest from each space.

Look at what you have in each layer. If your plan only calls for one or two different species in the ground cover layers, that may be a missed opportunity to add more diversity to your landscape and harvests.

Chicory root (*Cichorium intybus*) is a common roadside edible plant whose leaves are delicious in salads. The roots are also baked, crushed, and used as a coffee substitute.

Don't be afraid to mix different plants together within each layer to create a quilted landscape of plants that intermingle beautifully. This is a view of my garden.

What Is the Edible Part?

It can be easy for an edible landscape to lean too heavily toward one type of food. Fruits and berries are wonderful, but if that's what nearly all of your plants are producing, consider adding plants that can be harvested and used as green vegetables.

Don't overlook the many plants that can be used for teas, herbs, and flavorings. These are easy ways to add a little homegrown flare to nearly every meal. Even a quick reheated frozen dinner after a long day of work can feel like something a little special with a handful of fresh herbs from the garden sprinkled over it.

Finally, look at the list of what you plan on harvesting and make sure it matches what you like to eat. If you're really a coffee person, that tea hedge may be a good idea that you never actually use.

Usefulness is something you should revisit regularly. Many of the plants I recommend for your edible garden are simply not available in the grocery store, so you may have never eaten them until you grow them yourself. If you try something and don't like it, by all means replace it with something else.

NEXT PAGE: This garden (also shown on the book's cover) is a perfect example of a layered edible garden. It contains plants from every layer, from pear trees down to beetroots.

What Time of Year Is the Harvest?

This is a big one, and easy to overlook. Mark your spreadsheet with the time of year each plant will produce a harvest of food. With proper planning, you can make sure you have something to eat nearly every month of the year. Certainly spring and fall can be productive in any nearly climate.

If you live somewhere with cold, snowy winters or very hot, dry summers, your plants may be less productive, though in most cases you can still harvest at least something in those seasons.

Looking at your plant list, you may find that everything is producing just in the fall. This is often true for home orchards that include apples that, depending on the varieties, aren't ready for harvest until quite late. A bumper crop of apples in the fall is wonderful, but you need to eat all year. Round out your design with selections of food for other seasons.

Timing all your harvests without balancing the different seasons could mean that one instance of bad weather will knock out all your food for a year.

I made this mistake once with my raspberries. I had a big patch of all summer-bearing Boyne raspberries (*Rubus idaeus*) one year. We got hit with an early heat wave that basically cooked the developing fruit on the plant. What a shame, as summer-bearing varieties only fruit on second-year floricanes, meaning that after waiting over a year for that fruit, I had nothing to harvest that year at all.

You can also deal with seasonal variation in harvests by exploring how well your food can be preserved. You may not mind that you have an early summer glut of berries, because they can so easily go into the freezer and be enjoyed over the rest of the year. Delicate salad greens, on the other hand, really are best eaten fresh, so there's no point in growing more than you can eat.

This is an aspect of your garden design to revisit over the years. Notice what you are eating and enjoying, what you wish you had more of, and what you wind up giving away or just not harvesting. Let your edible landscape evolve with your tastes and diet.

An autumn harvest of apples, persimmons, figs, grapes, mushrooms, pomegranates, chestnuts, walnuts, and hickory nuts from a backyard edible landscape.

ORNAMENTAL FEATURES

The main focus of the edible landscape is, naturally, food, but we want it to be attractive as well. The plants recommended in this book are beautiful and also have edible parts, and there is no reason you can't have both growing in the same space. There are some simple concepts from design that you can bring to your garden to help make it a beautiful space, support a cohesive balance between your new plantings and existing ornamental features, and set the tone, feel, and style you want. You can do this, all while planning your harvest.

Color

The simplest way to make a garden look cohesive and well designed is to limit your color scheme. There are talented garden designers who can mix and match all the colors of the rainbow into one bed and make it look incredible, but if you're a beginner or less confident in your design instincts, simply limiting your color scheme is an easy shortcut.

You can find inspiration for a color scheme from a number of places. I got inspired by something as simple as the color of my house. With its dark grey paint scheme, it works well with the lighter-colored foliage or flowers I've planted in front. The combination creates a bold contrast. I currently have two shiny-silvery goumi shrubs that shimmer and look very interesting up against the dark exterior wall—and they give me edible fruit as well.

One way to choose a color scheme is to go to the nursery, find two or three of your favorite plants that look good when you set them next to each other, then only select additional plants in the same colors. Or just pick a color and ban it: No yellow flowers or foliage in this part of the garden.

You might be surprised that color matters as much in edible gardens as it does in nonedible ones. A rainbow of plants is available with colorful, edible foliage and plenty of plants with brilliant-hued berries and fruits that can either look beautiful together—or clash terribly.

Having a space for every color in the garden is also good for maximizing pollinator visits. Bees tend to prefer purples and blues, while hummingbirds will primarily visit hot red and orange flowers. Make spaces in the garden for each color to keep your design cohesive while also welcoming different pollinators into the space.

Yes, color matters in edible gardens too. There are many food plants with colorful flowers, foliage, and fruits (think of the many, varied colors of lettuces, chard, and other greens). Combine them wisely to create a garden that is visually appealing.

DEALING WITH IMPULSE PURCHASES

One of the difficulties of choosing a color scheme is the moment you fall in love with a plant that's perfect—beautiful, gives edible fruit, thrives in your conditions—but it isn't the right color. When this happens, one trick is to divide your garden space into different areas with different color schemes. You could put all your yellows and purples in the front garden and all the other colors in the back. That way, every impulse purchase has a place to go without interrupting your set color scheme.

This shade garden contains variegated Solomon's seal and ostrich ferns, both of which are springtime edibles.

Notice seasonal lulls in your garden, when not much visual interest is happening. Make note of it in a garden journal and then be sure to make additions to your garden in future years to fill in the blanks.

Don't just think about color with the plants in sunny areas of the garden. Shaded areas will generally be more dominated by foliage than flowers, but color still makes a huge difference. This is especially true of light colors, which will brighten a dark, shaded space.

Consider ostrich ferns (*Matteuccia struthiopteris*) or white variegated Solomon's seal (*Polygonatum odoratum* 'Variegatum'), which will brighten the area. Both can be harvested and eaten in the spring as well.

Texture

A plant's colorful flowers will be the first thing you'll see when you shop for plants. But flowers are fleeting, a surface treatment over the bones of your garden. Foliage texture is usually broken up into fine texture—think delicate fiddlehead ferns (*Matteuccia struthiopteris*) and carrot tops, bold texture—think big, broadleaved hostas (*Hosta* spp.) or rhubarb, and, falling somewhere in between, medium texture—think sorrel (*Rumex sanguineus*) or bee balm (*Mondarda didyma*).

While the rule of thumb for color is to limit your selections to a simple palette, the shortcut to using texture well is the opposite: Mix different textures together in every garden space. Including nothing but fine- or medium-textured plants is a common mistake, which results in a garden that looks a bit flat and uninteresting.

Look over your list of edible plants and make sure you have at least a couple bold-textured and a couple fine-textured selections in each garden area. This will help keep your design looking vibrant and beautiful.

TIMING

A classic beginning gardening mistake is to do all your plant shopping in the spring. You go, look at all the beautiful plants, and buy everything with a flower on it. Come summer, and fall, and winter, there's not much to look at as the plants wind down and go dormant.

While making your list of plants, note when each will look its best as well as when they produce. And that doesn't just mean flowering. It means flowers, berries and other fruit, attractive foliage, bark, fall color, and so on. Aim for a balance of interest, beauty, and productivity in every season of the year.

When it comes to arranging plants in the garden, there are a couple ways to approach seasonal interest. You can mix it all up, making sure every bed and part of the garden has something that looks great every season of the year, while another approach is to cluster seasonal highlights together. So maybe one bed looks fantastic in the spring, but then is quiet the rest of the year while another bed takes the spotlight in summer.

As your garden matures, you'll probably notice quiet times of the year when nothing much is showing off and getting you excited. A great way to fill in those seasonal lulls is to take a trip to a local public garden and see what looks beautiful at the moment. Note the names and make sure to add them to your own garden when the season is right.

OPPOSITE: This texture-filled garden offers a variety of edible plants, including gourds and squash on the trellis, and kale, chard, kohlrabi, and even taro root (*Colocasia esculenta*) at the ground level. The introduction of a selection of ornamental plants adds even more texture.

Problem Plants

One of the first things you should tackle in your garden is problem plants, often called invasive species. Invasive is a word that gets used a few different ways, so let's be clear about what this means.

People call some plants invasive because they can spread aggressively in a garden. But ecologists call a plant invasive when it spreads—almost always by seed—into natural or uncultivated areas. As they do, they can disrupt natural ecosystems by outcompeting native plant species for space and resources or by changing soil and moisture levels with their root systems.

Because invasive plants generally spread to new areas by seed, you often don't see evidence of their invasive potential in your own garden space. A plant may seem well behaved in your garden, not spreading out of its spot. And you may not see seedlings if you use thick plantings and mulch to minimize weed problems. But that plant can still be germinating and spreading where the seeds are being carried by birds or the wind into natural areas that you can't see.

Invasiveness is climate and region specific. Interestingly, many popular houseplants are considered invasive species in tropical areas, but as long as you live somewhere that freezes in the winter, there is no chance of them becoming invasive for you.

The internet makes it easy to find regional information about potentially invasive plants in your area. Each region's government usually has an agency that compiles an official list of invasive species. Look for your own area's list and become familiar with the common problem plants. When in doubt, opt out of a potentially invasive plant. There are many others to choose from, and there's no reason to plant something that could harm the natural ecosystem.

Other potential problem plants may depend on who you share your life and garden with. If you have a lot of small children running around, plants with large spines or thorns should be avoided, or at least carefully sited to the back of beds where there is less chance of accidental encounters. You could also incorporate such plants into a security hedge around the property to deter unwanted visitors (human or wildlife). Agaves (*Agave tequilana*) are beautiful and edible (tequila, anyone?), but they're so viciously armed that they aren't the most kid-friendly option for most gardens.

Agave plants are unique and edible, but they are also sharp and dangerous. Choose your plants wisely!

Problem plants that reseed overzealously (like this bronze fennel, for example) should be avoided if the plant is listed as an invasive in your region.

The many tiers of edible plants in this garden are evident. From the large, established apple tree, to the perennial layers of asparagus and rhubarb, and on down to the ground cover layer of strawberries and herbs, it is a space that is beautiful when viewed from either the kitchen window or the back patio.

TOXIC PLANTS

Toxic plants are often mentioned as a concern when planning a garden, especially around children and curious dogs. In fact, nearly any plant, even ones with edible parts, can be toxic if the wrong part is eaten. Good old potatoes and tomatoes have poisonous leaves. Any child who will enjoy your landscape needs to understand that they cannot eat anything in the garden without checking with you to ensure that it is indeed edible. They should also know that harvesting food from plants depends on knowing the right time to pick and then properly preparing the fruit, vegetable, or herb.

PLANTS PRONE TO PROBLEMS

Another group of plants with issues are those that consistently have pest or disease problems, require more or less water than you can provide, or otherwise aren't a good fit for your climate and garden conditions. Some of these problems will be apparent as you shop for plants, so can be avoided, but others will only make themselves apparent over time. If a plant is more hassle than it's worth, remove it! Taking plants like this out and replacing them with easy-care alternatives will make your gardening life much easier.

SCALE

Considering the height of your plants is key to good garden design and creating an effective, ecologically active, layered garden. Visually, varied heights make a garden dynamic and beautiful. Taller plants screen unwanted views while shorter plants fill in smaller spaces with beauty and food. Ecologically, different native insects, birds, and other wildlife thrive in different layers and heights, so keeping this varied will help enrich your garden's biodiversity.

The general rule is to put taller plants in the back of plantings and shorter ones in the front so you can see them all. This is a rule you can certainly break, especially with tall plants that don't grow densely, making it easier to see through them to the garden beyond.

Figure out what height plants you need by starting at the back of each bed and deciding what you want to see. If you have a view you want to screen out, choose plants tall enough to do that. If you want to be able to see more of the garden or a nice view beyond, choose plants that will come below eye level.

Don't forget to check those heights from inside the house, too, so you don't block views from windows. Then plan on tiers of plants getting progressively shorter, working through the bed toward the front. Low, ground-covering plants can be mixed in throughout to add extra insurance against weed invasions.

ACCESS FOR HARVEST

One of the big differences between designing an edible layered garden and one that is purely ornamental is that you need to be able to harvest. For each plant you'll include, note the edible parts and when you'll be harvesting them, then be sure to add access paths to be able to reach that plant at the proper time. Also consider how harvesting will affect the look of a garden space.

If you're going to be cutting many leaves or berries from a plant, consider adding other plants near it to fill in that space that will look bare after harvest. In balancing the ornamental and edible aspects of your landscape, you may also want to harvest a limited amount, leaving some fruit or other features in the garden to keep looking decorative.

If planted carefully, layered edible gardens are perfect places for children to roam and forage on their own, as long as you take the time to educate them on what is edible fresh off the plant and what needs to be cooked or otherwise prepared first.

AVAILABILITY

Your dream list of plants will always be tempered by what you can actually find for sale. If you're lucky enough to have a good independent garden center or specialty nursery in your area, you'll have more great options, though it may become a quest that takes you out of your way. Many gardeners have to drive far to find more than the generic selections available at the garden sections of the ubiquitous big-box stores.

The determined gardener, however, need not be limited to what they can find for sale locally. The internet has made finding mail-order nursery sources for unusual plants easier than ever. With a little time spent on your favorite search engine, you should be able to find sources for nearly anything that you want.

Be aware, however, that plants ordered online and shipped to you will almost always be smaller and more expensive than the same plants from a local nursery, and they will sometimes arrive stressed from the shipping process. No plant has evolved to thrive sealed in a box bouncing around in a truck, so unpack your orders right away to help them start recovering. Plants shipped when dormant, without leaves, will generally go through shipping easier than those in active growth.

Bare-root trees are generally less expensive than potted trees. They are also easier to transport and plant. Bare-root planting is an excellent option for most fruit and nut trees.

PROPAGATION

Basic plant propagation can help expand your garden. By taking cuttings, dividing, or growing a special plant you can't find for sale from seed, you can easily turn your one expensive, mail-order plant into a dozen to plant around your garden through vegetative propagation. This may sound complicated, but it really isn't, if you understand the rudimentary technique and have the right equipment.

I'm lucky enough to have a greenhouse, which makes starting seeds early for my own garden a breeze, but an expensive greenhouse certainly isn't necessary to grow beautiful plants for your garden. It used to be that artificial lighting for starting seeds or rooting cuttings indoors was expensive, both to purchase and run. With the recent availability of affordable, energy-efficient LED lights, creating greenhouse conditions is easy and affordable, and anyone who wants can set up growing conditions indoors.

Propagating from Seed

Growing from seed is perhaps the most cost-effective way to get many interesting edible plants for your garden. Because seeds are small and ship well, you can easily order from a wide range of different sources, getting more varied, unusual plants than you would ever be able to find at a local nursery.

Seeds are also easily collected from plants you already have in your garden or (with permission) from a friend's garden. Seeds are nature's propagation unit: baby plants with a starter food supply, sealed in a tiny, durable package. Germinating them and growing them on to mature plants can be a little fussy and requires patience, but it's not as difficult as you might think.

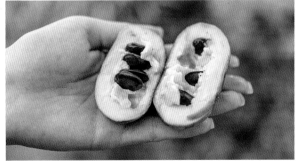

Seed propagation from edible plants like this pawpaw (see page 119) is fun, but you do need to investigate the best techniques for each species.

Starting Outdoors

The simplest way to start seeds is to plant them outside and let nature do all the work for you. You'll get less than 100 percent success with this method, since nature isn't always kind to baby plants, but trying a few tricks will give your seedlings a fighting chance.

If you're a beginner, you can try focusing on large seeds. With more stored food, the emerging plant can get more nutrients and emerge as a bigger and less fragile plant. Tiny seeds, on the other hand, produce tiny seedlings that are more delicate, making surviving the early stages of life more challenging. Beans, squash, sunflowers, and even trees like oaks (*Quercus* spp.) start out as large, robust seeds that develop into large, robust seedlings.

When you've selected your seeds, plant them where you want the mature plants to grow. Be sure that the temperature, light, and moisture in the location are appropriate for the kind of plants you're starting, and keep the space watered and weeded as the seedlings germinate and establish themselves.

You can compromise on this method by sowing your seeds in outdoor containers. Place your containers in a sheltered spot where the seeds won't be washed away by heavy rain or blown by winds, and consider covering them with something to keep out hungry birds and pests.

Many gardeners like to use clear plastic containers with the tops cut off to hold their seedlings. You can also make covers out of old window screens to allow light and heat in while excluding pests and weed seeds. This is an easy way to baby small, delicate seedlings. If you use this method, make sure you vent the lids on sunny days to prevent plants from cooking under cover, and check on moisture levels to prevent seeds from drying out. Newly germinated seedlings that dry out are goners.

A friend loves to start seeds of her favorite nut trees in fabric pots outdoors at the same time she starts seeds of shrubs and perennials.

Many annual and herbaceous perennial edible plants are very easy to start from seed, including this crop of basil seedlings which will soon be separated and transplanted into larger pots.

Starting Seedlings Indoors

Whether you have a greenhouse or are growing under lights, starting seeds indoors uses a process similar to starting them outside.

You can maximize your success and use your indoor space effectively by sowing in seed pots. This involves sowing many seeds together in one larger pot, instead of sowing one or two seeds in a small pot. Once they've germinated and are big enough for you to handle, slide the whole mass of seedlings out of their pot and tease them apart.

Don't be afraid that you'll damage the seedlings when you pick them up. You can avoid hurting them by gently grabbing each by its cotyledons (seed leaves), not the very delicate hypocotyl (the lower "stem" portion). If a cotyledon tears, a seedling can keep growing, but if the hypocotyl breaks, it won't be viable.

Gently tease the roots and shake until the soil falls away and you can pull them apart. There's a tendency to warn against letting the soil and media dry out a bit prior to separating seedlings, but, in my experience, this will make the job easier and lessen the chance of plant roots tearing and being damaged.

Wet media acts as a kind of glue and can make teasing apart individual plants difficult. I usually water thoroughly one or two days before I do this, then let them dry a bit before separating. This is my rule of thumb for finding a sweet spot between easily separating the roots successfully and attempting this when they're too dry and stressed.

You'll have better luck if you start your seeds in pots and separate them once they've germinated. This can save you from wasting seedlings or pots.

When you sow directly into the plants' final container, with one seed per pot, you may wind up with a few empty pots where the seeds have failed to germinate. Sowing two or more seeds per pot may force you to pinch off and kill the extra ones that sprout because they won't fit together in the container.

Planting in a seed pot and then dividing each seedling gives you one plant in each pot—no waste and no empty pots. Over time, as you gain more confidence, you can start seeds for similar crops together in a large flat, as the bigger mass of soil doesn't dry out as fast and it's easier to move one flat around than it is to move multiple little containers.

Propagating from Cuttings

Working with cuttings is another great way to propagate a plant that you love or one a friend shares with you.

Plants can root from cuttings because plant cells are totipotent, meaning that any of their cells can, in the right conditions, divide and develop into any other type of plant cell. This is one of the key distinctions between plant cells and animal cells. Humans and other animals do not have totipotent cells—once they get beyond the early stages of embryonic development, animals' skin cells can only grow into more skin cells, muscle cells can only form more muscle cells, and so on.

Plants cells are different. Every cell in a plant is, essentially, an embryonic stem cell. When you take a piece of stem, the cells in that stem can divide and turn into root cells, growing into a whole root system.

That's the theory, at least. In reality, you need to give the cutting the right conditions to allow it to grow new roots and develop into a whole new plant.

Cuttings need two things: light and water. Light allows them to photosynthesize and produce the energy necessary to grow those new roots, and water keeps them from drying out so they don't die before the roots appear and allow them to take up water from the soil. Your cuttings will need nearly 100 percent humidity while their roots are growing. This is most critical with cuttings from soft, new growth, less so for more mature stems.

Commercial propagators use something called a mist bench to achieve this. It's a spot in the greenhouse with nozzles on a timer that spray down the cuttings with a fine mist every few minutes. This keeps them fully hydrated and alive while they grow new roots.

Most home gardeners don't have a whole misting system, as this setup can be costly, so the easiest solution is to use a clear plastic container that is mostly solid but slightly vented to allow a little air movement. You can buy premade containers for this purpose or reuse plastic containers you got to transport food from the market or for take-out from a restaurant; these usually have clear lids that can lock in moisture, achieving the same effect as you would get from misting.

Add your cuttings to moist soil in your containers, put the lids on slightly open to allow for venting, and place them somewhere with bright light but away from direct sun. (These points are important: The clear plastic lids can hold a lot of heat when sealed, and in direct sun your plants could quickly overheat and die.) With your containers prepared, place them under grow lights you might use for starting seeds, or a bright spot in the north side of the house works just as well.

You should also invest in a small container of rooting hormone if you're working with plant cuttings. These are synthetic versions of the natural hormones a plant produces to tell those stem cells to change into root cells. Rooting hormone is affordable, lasts a long time, and will give you a much better success rate with your cuttings. Dip the cut bottom end of each cutting into the rooting hormone powder or gel before sticking it into the soil.

Pro tip: A little rooting hormone goes a long way! Pour out a bit into a small saucer (whether you're using powder or gel) before dipping in a plant stem. Avoid dipping the stem directly into the container, which will prevent debris and water from entering the hormone material, and will also give you a better view of how much you've applied.

There are many recipes available online for making your own rooting hormone from willow twigs or other materials, but I don't recommend using any

A simple humidity dome made of half of a plastic soda bottle is enough to keep cuttings moist as they root.

How to Take a Stem Cutting

Taking a stem cutting is a fairly easy process.

1. For herbaceous edible perennials, tropical edible plants, and most annuals, cut off fresh new stems to create your cuttings. Select cuttings from new growth, as this will grow roots most easily. Avoid taking flower buds, if possible, because developing flower buds sap energy from the cutting. If you can't find any growth without a flower, pinch off all the flower buds. For most herbaceous stem cuttings, aim for four nodes, the visible bumps at the base of each leaf, from which the buds emerge.

2. Trim off the lowest leaf or two from the cutting. This is important: At the base of each leaf is a dormant bud, and at the center of the bud there are cells called the meristem. These undifferentiated cells will divide and develop into new stems. The cells of the meristems are the most capable of transforming into new root cells, so your new roots will most likely come from around these buds at the leaves' base. When cutting, I like to make a nice square cut, a half-inch (one centimeter) or so under the lowest node.

3. When processing cuttings, it's important to keep the cuttings moist. Don't allow your cuttings to dry out! Keep them cool, moist, and out of direct heat or sunlight until you're ready to dip them in rooting hormone.

4. Dip the base in rooting hormone to tell the meristems to start changing into roots, stick the cut end into a pot of sterile potting soil, cover it with a humidity dome of some sort, and place it somewhere with bright, indirect light.

Then you wait. Many perennials and annuals will root in a week or two, especially if they're kept warm. I recommend adding bottom heat via a standard propagation or seed-starting heat mat, as it can speed up the process greatly. Other plants will take longer. Camellia (tea) shrubs, for example, can take months to grow roots, but as long as they're kept moist, the roots will eventually form.

New to taking cuttings? Many culinary herbs in the mint (Lamiaceae) family like sage, basil, and thyme are forgiving and root readily. For most stem cuttings, leaving up to ½ inch (1 cm) under the lowest node ensures any die-back will not disrupt the rooting process.

of these. It isn't that they don't work—you can extract plant hormones from these natural materials—but it's difficult, time-consuming, not very effective, and may not give consistent and reliable results, which is something you want.

Going with a commercial product helps rule out "Was the DIY rooting hormone not strong enough?" when troubleshooting issues with cuttings. Commercial products will work better and are widely available, fairly standardized, and cheap, so I don't see a reason to make your own.

Not all plants can be rooted easily from cuttings. Generally, annuals and herbaceous perennials root the easiest, shrubs harder, and many trees are very difficult to root from cuttings.

Timing Matters

When you take your cutting matters. Soft wood cuttings are what the name says: They use fresh, new, flexible growth. Other plants root better when taken from semi-hardwood cuttings, using plants with more mature and firmer growth.

A quick internet search will often turn up tips for choosing the right time to take cuttings from a particular plant. Use the information you find, and you can always experiment. Take cuttings from the same plant at different times of the year and see which works the best. Don't be afraid to experiment here. It doesn't cost anything, and you'll quickly learn the best times to take your cuttings.

What to Do Next

There are a couple of ways to tell when the cuttings have rooted and are ready to transplant. Give the cuttings a gentle tug—if they resist, that means their roots have grown and taken hold. If they pull out easily, the roots haven't formed.

Be careful and patient. If you pull too hard, you can damage the new roots. If you disturb the cuttings too often, they'll be slower to grow roots. Tugging gently on the cutting is the fastest way to see if your plant has rooted—after you've done this a few times, you'll know how to feel the difference.

Another way to check for rooting is to gently scoop under the plant with your hand or a small trowel and lift it up to see if it has roots. If the media is moist (which it should be), you can tip the small nursery pot upside down into your hand, then gently lift the pot to see if roots have emerged. I use this method most often because it disturbs the cuttings the least.

Another good clue is to look for new growth. Actively growing roots produce a hormone that stimulates new shoots, so new growth pushing from the top of the cutting is nearly always a sign that roots are growing beneath.

Once you have some roots, you can remove the cutting from its high-humidity environment, but you should still baby it a little. New cuttings can be delicate, so make sure to water them well and keep them in light shade, slowly moving them out into more direct sun as they continue to grow roots.

Taking cuttings of common perennial herbs, such as this rosemary, is a fun project that requires very little investment of time.

Some edible plants, like these raspberries, layer on their own. When the tip of a cane contacts the soil, roots naturally form. You can speed this process by intentionally bending a raspberry cane down to the soil and pinning it in place until roots are formed where the plant touches the soil.

Division is another popular way to create more plants. Here, a gardener is dividing a hosta plant whose spring shoots are tender and delicious.

Propagating by Layering

Plants like edible shrubs are slower or more difficult to root. A great way to propagate plants like these is a process called layering. You can easily clone fruiting shrubs like gooseberries, elderberries, and currants, and cane plants such as blackberries and raspberries.

This is the most passive way to create more plants. At its simplest, this method for propagating involves bending the branch of a shrub down to the ground and putting a rock on it so the branch stays in place. After a few months, check the branch to see if it has grown roots where it touches the soil. If it has, cut the branch that connects it to the mother plant and move the plant to another part of the garden.

You'll find there are many variations on this basic process. To speed things up, scrape the bark away where it touches the ground and dust it with rooting hormone. You can also bend the branch down to touch a pot full of soil instead of the ground; this allows you to extract it more easily than having to dig it up once it roots.

Another method is called air layering, which brings the soil up to the branch rather than bending the branch down to the ground. Using aluminum foil makes this easy: Cut into the branch with a knife, dust the injury with rooting hormone, then use foil to set a mass of potting soil or sphagnum moss around the injured site. Be sure that this segment of the branch doesn't dry out. Roots will develop inside the foil within a month or two. You can also buy special tools to do this, but aluminum foil works just as well and costs a lot less.

Propagation by Plant Division

Division is one of the easiest ways to propagate herbaceous edible perennials like daylilies (*Hemerocallis* spp.), oregano (*Origanum vulgare*), chives (*Allium schoenoprasum*), and hostas. Many ground cover plants, such as thyme and bunchberry (*Cornus canadensis*), can also be propagated using this method.

It's the perfect way to save money while multiplying the quantity of ground-hugging plants that work well in your space. Division is as simple as digging up a clump of your favorite plant, pulling or cutting it into pieces, then replanting each section. The best time to do this is usually in the early spring, before any fresh, fragile stems have emerged, since you want to avoid damaging these parts. Bear in mind that some plants prefer to be divided in the fall, so do some research on your plant before trying to divide it.

Novice gardeners may find division intimidating, as it seems pretty radical, but in fact most plants will respond well to this process. They may even grow better for having been thinned out and spread around a little, as they won't have to compete with their neighbors for water, nutrients, and sunlight. As with all of these methods, keep the new divisions well watered while they settle into their new homes. Even if you don't need more plants, divisions are perfect gifts for your gardening friends.

SHOPPING FOR PLANTS LIKE A PRO

There are several ways to shop smart and get the best deals on the highest quality plants when looking for plants for your new garden.

Buy Small

This may seem counterintuitive, but it's best to buy the smallest plants you can. Large plants are tempting because they look so lush and full and can give an instant garden effect when planted, but there are downsides. Larger plants are more expensive, harder to transport, and much harder to plant.

When you have to get hundreds of plants into the ground, not having to dig enormous holes for each one is a huge benefit. And the difference in size won't be a problem once the small ones quickly grow up to their full mature size and look just the same as the more expensive, large options.

The only real downside to planting very small plant sizes is that they can dry out faster when newly planted. Irrigate your new plantings, of any size, for their first year—this is especially important for small plants.

Look for Good Root Systems

Far more important than what you can see above the pot is what is happening down below. In nature, plant roots fan out into the soil, seeking moisture and nutrients. In a pot, those roots hit the pot wall and start growing in circles, looking for a way out.

A plant that has been in a too-small pot for too long is said to be potbound, with roots circling inside the pot. When planted out, those roots have trouble changing course and moving out from the original area of the pot and into the surrounding soil. When shopping for plants, look for ones that don't have a lot of top growth compared to the size of the pot. A huge plant in a little pot usually has terribly potbound roots.

You can improve this situation by removing the outer layer of circling roots. There are a number of ways to do this. For small roots, you can pull away the circling roots on the sides and bottom of the soil mass with your fingers. For larger roots, use a sharp trowel or pruning shears to cut through the roots, or you might try shaving off the outer layer of roots from the sides and bottom with a pruning saw.

Healthy root systems are a must. If the roots are circling around inside the pot, be sure to loosen them with your fingers or a trowel before planting.

This may seem extreme, but don't be shy about this step! A healthy plant placed in its ideal home will recover and benefit from this step quickly. Those circling roots aren't doing the plant any good, and pruning them away will stimulate new roots to grow out into your garden soil, helping the plant establish itself much faster.

EXTRA CONSIDERATIONS FOR TREES

When shopping for trees, pay close attention to what you are considering, so that you can be sure you're buying healthy plants. A healthy, well-grown, well-planted tree will easily outlive you and be the most beautiful thing in your garden. A tree with serious problems can grow big enough to cause some damage and then fail spectacularly. Spend whatever time you need to get your trees right.

The key to looking for a healthy tree lies in its root flare. The root flare is that point where the trunk transitions into roots, where the roots flare out into the ground. Healthy trees should not have trunks that go straight into the ground like a telephone pole, rather they should show a visible fan of roots spreading out from the base.

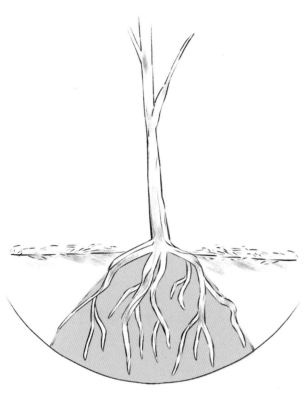

The root flare, or the shoulder where the base of the trunk meets the start of the roots, should perch 1 to 2 inches (2.5 to 5 cm) above finished grade. Young, bare root trees can be perched atop a firm mound of soil to help hold the position while backfilling to secure the tree. This method also ensures roots grow outward to avoid girdling.

The problem is that baby trees, when grown in pots, are easy to plant too deep. Look for that flare of roots at the base of the trunk when you are shopping. It should be visible just at the soil surface. Trees planted with their root flare buried are vulnerable to what are called girdling roots. As the tree grows after being planted too deeply, the roots circle around the edge of the pot, making a loop around the trunk. The tree will grow and the trunk and roots will get bigger, with that loop of roots turning into a noose, cutting off the flow of water and nutrients from the roots and eventually killing the tree.

This is a terrible problem because, while the tree is small, everything seems fine. The girdling roots won't appear as a problem until the tree is large enough to do some damage, after which the condition will kills the tree.

With any tree you buy, edible or otherwise, look for that root flare. If it isn't visible, dig down in the pot and remove any soil and roots above the natural root flare before planting, then plant the tree with the root flare perched just a bit above the soil surface.

This may seem too high, but soil settling over time will bring the tree lower. The extra bit of height also keeps mulch—whether applied or naturally occurring from fallen leaves—from building up and resmothering the root flare area, which can also lead to pest and disease issues.

PLANTING TIME

When exactly is the best time to plant your layered edible garden? Most people assume that spring is the prime planting time. And, for many climates, it is.

The truth is, the best time to plant depends on your local climate. There are pros and cons to planting in the various seasons.

Spring

In most climates, spring is when most planting takes place. The nurseries and garden centers will have the most plants in stock. If you live in a climate with long, cold winters, there's an energy and excitement during the spring season that's infectious and makes you want to plant. Enjoy that excitement, but remember a few factors when putting plants in the ground in the spring.

Unpredictable freezing temperatures and frost define the spring garden. You should plant after the last chance for a frost. A quick search online will tell you when the last frost-free date is for your area.

Some plants can grow happily through a freeze. When planting plants from the nursery, however, it's best to err on the side of caution. Plants from the nursery have usually been growing in a greenhouse and are acclimated to warmer temperatures. These plants will be more sensitive to a late freeze than the same plant if it had been growing outside. You don't want

Investigate Insecticide Use at the Nursery

In the layered edible garden, we want our plants to provide us with food, as well as supporting the insects and other wildlife around us. Do your homework on the insecticides that might have been used on the plants you are buying.

This is critical if you're buying something that is edible but not one of the traditional vegetable species. Lettuce and hostas are edible greens, but growers can legally use many, far more toxic chemicals on a hosta than they can on a lettuce plant meant for eating.

Good growers, like specialty nurseries and independent garden centers, should be able to tell you exactly what was sprayed on the plants. You can look up the products they used and see how toxic they may be and how long they will persist on the plant. See if the insecticide is approved for use on vegetables; if it isn't, that's a red flag. If you can't find any information on the chemicals that were used, a good rule of thumb is to wait a year before harvesting from a new plant to ensure any insecticides have fully worn off.

As bees forage on flowering shrubs, whether they are edible, like this honeyberry/haskap (*Lonicera caerulea*), or not, they are exposed to any pesticides used on that plant. Ask questions when you purchase plants so you know what you are introducing into your garden.

to run the risk of putting your hard-to-find plant in the ground and then having emerging buds and leaves damaged by sudden temperature dips. Be patient and hold out for consistent warm, frost-free weather.

In climates with snow, where the soil freezes in the winter, early spring is characterized by wet, muddy soils. If you have heavy clay soil, you should avoid digging, planting, or even walking on the soil until it has had a chance to drain and dry out a little. Disturbing damp soil while it's completely saturated will destroy the soil structure and lead to soil compaction. If you can dig up soil and form it into a mud ball, it's too wet. Wait until it has dried enough for a handful of soil to crumble when you release it.

Summer

Summer plantings can work if you live somewhere with cool temperatures and regular rainfall. The hotter and drier your summers, the more challenging you'll find keeping new plants alive while they settle in and spread their root systems. If you need to water your lawn in the summer to keep it green, then it's probably too dry in your area to plant in the summer. Better to plan, wait, prep the soil, and plant in the fall.

Fall

Fall plantings can be a wonderful option, and this is actually my favorite time to plant trees and shrubs. One of the best perks of planting in this season is that many nurseries have sales to clear out their inventory before the winter, so you can get plants at a discount.

If you have a hot, dry summer, fall plantings are perfect, because they give your plants the maximum time to grow and establish a deep root system before hitting next summer's dry period.

Using Plant Growth Regulators

Another group of chemical sprays is widely used in nurseries but is much less talked about: plant growth regulators (PGRs). These are chemicals that imitate the effect of various natural plant hormones and are used to keep plants from growing too tall or too fast at the nursery. Growers for the big-box stores use them extensively to keep plants short enough to fit in the semi trucks that transport them to the store, but they are common in a lot of independent nurseries, too, especially with annuals. You should be aware of them for two reasons.

First, if overapplied, plant growth regulators can effectively stop a plant from growing any taller. If you bought a bunch of tomato transplants that sit around without growing, you might have bought plants that were overtreated.

Second, many perennials will look different, and be much taller, once they've settled into the garden and the plant growth regulators have worn off. Just because a hibiscus is blooming at 6 inches (16.25 cm) tall at the nursery doesn't mean you'll see the same result the following year in your garden. To avoid such surprises, be sure to read the label.

Finally, and most critically, most plant growth regulators are not approved for use on edible plants, which means that they are not considered safe to eat. Again, when planting an edible landscape, ask the nursery what has been sprayed on your plants. If you have any questions, wait to harvest for a year after planting to ensure that any chemical treatments have worn off and a new cycle of growth has emerged for you to harvest.

Fall is a less ideal time for planting in areas with very cold winters. If your soil quickly freezes deep, fall plantings will not have time to establish their roots and can be pushed out of the ground by frost heaves.

Don't plant anything in the fall that is on the edge of winter hardiness, as they will only develop cold tolerance after they've had a full growing season to get established. If a late planting is your only option, be prepared to protect both the upper and lower portions of the plant before temperatures drop. Gather frost blankets, burlap sacks, and mulch to help buffer the new plant from the chill.

Winter

You might not think of planting in winter—and certainly it is impossible in cold, snowy climates—but in milder climates where the soil doesn't freeze during winter, you can consider planting during this season. Be sure that the plants you're putting in the ground have already been growing outside so that they are used to the cold. Even if your plant is very cold hardy, moving it from a warm greenhouse straight into a winter garden will kill it.

But if you can get plants that have been outside and acclimated to winter temperatures, and your soil isn't frozen, the cool months of winter can actually be a good time for plants to settle in and grow roots in preparation for warmer, dryer days ahead.

YOU'RE READY FOR THE NEXT STEP

With the information in this chapter, you have everything you need to compose and plant your perfect edible landscape. Taking the time to compose your perfect landscape and source the best, healthiest plants will pay huge dividends in your garden for years to come.

Next, let's take a closer look at the eight layers of the layered edible garden and meet some of the best plants from each one.

NEXT PAGE: Following proper planting techniques is a must for setting yourself—and your garden—up for success.

4

THE LAYERS

Now it's time to dive deeper into the eight plant layers that create a layered edible garden and meet some of the best plants within each group.

WHILE YOU MAY ONLY HAVE one or two plants in the largest layers (canopy trees and subcanopy) depending on the size of your space, you'll likely have more specimens of each smaller layer. Each layer has a role to play in the design of the garden, in what's available for eating, and in the ecology and biodiversity of your space. When selecting plants, consider all the factors I outlined in the previous chapter and refer to your spreadsheet of traits for each of your wish list plants.

As a reminder, here are the eight layers of the edible landscape:

Canopy trees
Subcanopy trees
Shrubs
Herbaceous perennials
Climbers
Annuals
Ground cover
Root crops / rhizosphere

Incorporating as many layers as possible into your garden will result in the most productive, diverse, and resilient garden possible.

The Canopy Tree Layer

The biggest, most dominant layer of your garden will be the canopy trees. These are the large trees, reaching over 40 feet (12 m) tall, making them the layer that casts the most shade and has the greatest impact on the rest of your garden space.

A large canopy tree will probably be the single most important plant in your landscape, but you won't have a great deal of choice when it comes to this layer of your edible garden. Trees grow slowly and live a long time, at least when planted properly. You'll usually inherit what someone else, perhaps nature, planted in your space and have to work with what that tree gives you.

Before you make a final decision about keeping what you have or planting new trees, evaluate the existing trees on your property and ensure that they are healthy and well placed. It's always heartbreaking to take down a mature tree, but sometimes it has to be done. Removing a tree is sad, but, especially on small, urban properties, it's important to take care of a potential hazard before the tree topples over onto your house or a neighbor's garage.

There are a few things to look for when examining your trees and seeing if they have the potential to be a problem.

Root flare. As discussed in the section on purchasing trees, ones that have been planted too deep may be doomed. If its trunk goes straight into the ground, you may be able to save the tree by digging down to find the true root flare, then cutting off any circling roots you find. But if the tree has been growing too long as it is, this may be impossible. Improperly planted trees, with circling roots that will eventually girdle the trunk, are ticking time bombs and should be assessed by an arborist to determine the need for replacement.

The tree layer contains the largest and longest-lived edible plants. It is an important one to get right from the start.

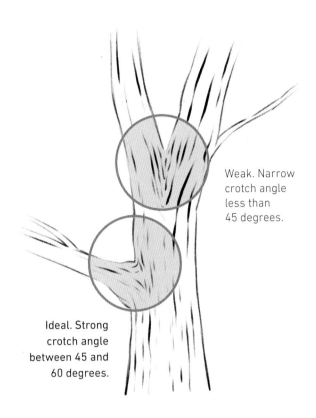

Weak. Narrow crotch angle less than 45 degrees.

Ideal. Strong crotch angle between 45 and 60 degrees.

Narrow crotch angles. Crotch angles represent the angle at which branches attach to the main trunk of the tree. The ideal angle is between 45 and 60 degrees, while angles smaller than 45 degrees are points of weakness prone to disease and damage.

Once hugely popular street trees, Bradford pears (*Pyrus calleryana*) are infamous for having tight upright growth that often collapses dramatically as the tree gets older.

If most of a tree's branches are fine, but a large one hanging over your house has a bad crotch angle, you can probably just remove that branch and preserve the rest of the tree. This is most likely a job for an arborist to ensure the branch is removed safely, without damaging the tree.

Diseased or damaged trees. Problems may be obvious if a tree is ailing: many dead branches, serious damage on the trunk, or a leaning trunk. You're probably better off having the tree removed if you see any of these issues.

If you aren't sure, it may be worth having an arborist inspect the tree. Be certain to hire a properly trained, licensed arborist. There are unscrupulous tree companies that advocate for the removal of healthy trees or that sell unnecessary pruning or fertilization treatments. A good arborist will be able to evaluate trees and give you helpful advice, as well as make sure any removal is performed in accordance with local laws and safety standards; some municipalities require a permit to remove a tree.

When deciding if a tree needs to be removed, take the location into account. Some issues are obvious, such as if the tree or its branches are hanging over a power line, house, or other structure. On larger properties, trees that won't do serious damage if they fall can be left in place. Many trees that seem damaged can actually still stand for decades and provide a habitat for birds and insects. Or they might fall in the next big storm. Evaluate how much damage a falling tree will do and adjust your risk assessment accordingly.

Working with the Trees You Have

The healthy trees on your property will have a huge impact on your decision for how the rest of your edible landscape works. Trees create the most shade, provide free mulch in the form of fallen leaves, and focus attention with their wonder and beauty. Trees can also make parts of your garden significantly drier thanks to their roots, and their large bulk will block—or frame—the views from your garden to the world around you. Your big tree's cycle of leafing out, flowering, and going dormant in the winter will also define the seasonal rhythm and feel of your garden space.

The dry shade created by thirsty tree roots can be hard to manage. Some species, such as maples (*Acer* spp.), are difficult to garden under as their roots suck up every drop of water, even if you irrigate regularly.

While there are plants that will grow in these conditions, your best choice here is often a container. Fill a large pot with good soil, place it under the tree, and you can grow all kinds of things without competition from the tree roots. Be aware that, when you place a container directly on the ground, tree roots may grow up the drainage holes into fertile, moist soil inside. To avoid this, set your container on some sort of small support—pot feet, bricks, even a rock will work—just enough to keep it from touching the ground; this will keep the water you give your plants safe and for their roots only. Be sure to not go overboard with this; tree roots need water and air to move into the soil to stay healthy, and covering large areas with containers or beds can harm them.

Oak (*Quercus* species) is a favorite tree to include in an edible landscape. See page 114 for more about this valuable tree species.

Sugar maples (see page 111) can be tapped for their sweet sap. They also offer stunning fall color and serve as a resource for many insects and other animals.

Don't ignore the edible aspects of your existing trees. Even if you didn't choose a tree, you may be able to harvest food from it.

Acorns from oak trees can be eaten if processed properly. Acorn flour noodles are beloved in Korea, sugar maples (*Acer saccharum*) can be tapped for their sap, and the new growth of many conifers like spruces and pines can be used to make tea or other beverages. Since the tree is already there, you might as well take advantage of it.

The first step is to identify the precise species of trees on your property. Local field guides are great for identifying trees. Plant identification apps can be helpful but should be taken only as a starting point for correct identification, rather than as the final word; they can misidentify species, especially less common ones. Once you know exactly what you have, do your research to find out if your tree has edible parts and how best to prepare them.

Adding New Trees

Once you've taken stock of the trees you have and are keeping, you may find that you have space to add more. Planting a tree is one of the great joys of life, but you shouldn't rush into this. Take more time researching, buying, and planting a canopy tree than anything else in your garden; it won't be you alone, but future generations, who will live with the consequences of your decision. You don't want to plant a tree that someone else will just have to remove because it's in the wrong spot.

There are a few good choices to start with listed on page 110. If possible, find mature specimens in local gardens or arboretums to see what will happen once your baby tree matures. You should also consider a tree that will bring edible benefits to your garden and, if possible, sample the food that your potential tree will produce. For example, taste a black walnut (*Juglans nigra*) before planting this tree, so you know if you'll enjoy eating its fruit. See how harvesting and preparing the edible parts of your tree is done. Nearly everyone loves maple syrup, but are you going to want to go through the long, somewhat sticky process of turning maple sap into syrup?

In addition to the positive aesthetic and edible aspects of a potential tree, don't forget to consider possible downsides. Research the susceptibility of trees to pests and diseases, especially in your region. Some species of trees will be a great, low-maintenance option in one climate, but they can be disease magnets if grown outside the temperature range and seasonal moisture they're not suited for. There are regional pest problems as well, especially as new invasive pest species are introduced.

Some edible trees can be quite "messy." Take that into consideration when making your choice.

Consider whether your potential tree will drop a lot of unwanted material all over the garden. Fallen leaves can make a great mulch, but some species are messier than others. River birches (*Betula nigra*) are a popular example. Though birches produce edible sap that can be cooked into syrup, they tend to drop twigs throughout the year, requiring constant cleanup. This will be more of a concern in a small, intensively gardened yard, and less so in a bigger, wilder space.

If you're planting close to the house or other buildings, make sure the tree isn't known to be weak wooded and break up in storms. Where I garden, Douglas firs (*Pseudotsuga menziesii*) are a beautiful native tree, but they're notorious for dropping huge limbs in windstorms.

Finally, consider the potential irrigation needs of a tree. If it requires constantly moist soils, and your climate doesn't provide that, you'll probably be better off choosing a more dry-adapted species.

Shopping for Trees

Tree shopping is also the perfect time to visit that special nursery with the widest selection and healthiest specimens. Be sure to ask the employees what they'd recommend. Specialty growers love their plants and are a great resource for what will do well in your local climate. This is the biggest, longest-lived investment you'll make in your garden, so don't settle for whatever is cheap and available at the closest big-box store.

When shopping, choose field-grown trees over those grown in pots their whole lives. Also known as balled and burlap trees, they will be sold with their root ball wrapped in burlap. Because they haven't been grown in pots, you don't have to worry about circling roots and the many problems they can cause. If you do opt for a container-grown tree, try to find one that has been grown in soft fabric pots or special containers that have holes in their sides, which allow air to move through the pot and keep roots from circling the inside of the pot. The traditional, solid plastic pot is your worst option, and you should plan to prune off the outer circling roots before planting if that is all you can find.

However you buy your tree, look for a visible root flare at the base of the trunk and a single, straight, nonbranching trunk without any narrow crotch angles. Remember to water regularly and deeply for the first year or two of the tree's life while it establishes its root system.

PINE

Pinus species

Layer: Canopy tree

Edibility: Needles, immature cones

Light requirements: Sun

Size: To 30 or more feet (9 m), depending on variety

Habit and form: Evergreen tree

Flowering and fruiting times: NA

Native range: Worldwide

Lowest hardiness temperature: −40°F/−40°C

Ecological benefit: Food source for many mammals

Pest and disease: None serious

Propagation: Grafting or seed

Pollination requirements: NA

Pine trees are a familiar part of nearly any landscape. Their tall, evergreen habit makes them perfect privacy screens. They are particularly good for deadening road and other noise.

You may be surprised to learn that you can eat them: Nearly every part of the tree is edible. The needles are most often eaten, especially on tender new growth, as well as the immature cones. Pine needles are most often used as a flavoring, steeped in hot water to make tea. Immature cones, while still tiny and green, can be simmered in water and sugar until tender, then the liquid is reduced to a thick syrup to make pine cone jam. The needles and cones of spruces (*Picea* species) can be eaten the same way.

↑↑ Foraging for spruce tips. ↖ Harvested spruce tips ready for preparation.
↗ Young pine cones can be made into a delectable jam.

SUGAR MAPLE

Acer saccharinum

Layer: Canopy tree

Edibility: Sap

Light requirements: Sun

Size: To 50 feet (15 m)

Habit and form: Deciduous tree

Flowering and fruiting times: NA

Native range: Eastern North America

Lowest hardiness temperature: −40°F/−40°C

Ecological benefit: Host plant for many butterflies and moths

Pest and disease: None serious

Propagation: Seed

Pollination requirements: NA

Sugar maples are shade and street trees beloved for their beautiful spreading crown and brilliant fall color. They produce a sap that is highly prized as the basis for maple syrup.

Making your own syrup from a maple tree in your yard isn't difficult, though it can be time-consuming. In late winter and early spring, when temperatures rise above and then drop below freezing at night, the sugary sap will flow up from the roots to the buds to stimulate their first flush of growth. Once the buds begin to appear with consistent warm weather, the sap flow season is over.

You can buy taps online. These are hammered into the tree and then buckets are used to catch the sap. A single mature tree can produce 10 to 20 gallons (38 to 76 L) of sap in a season, which sounds like a lot, but it takes 40 gallons (151 L) of sap to make one gallon (3.8 L) of syrup. Sap is mostly water, so it has to be boiled down to remove the excess water and increase the sugar concentration.

Take all the sap as it fills the buckets, put it in a large pot over a burner and boil it away, driving off the water to concentrate the syrup. This is the most important part: *Boil down the sap outside*. If you attempt this in your kitchen, you will have a sticky layer of sap covering everything. An outdoor grill or portable electric induction burner work well for the boiling-down process.

You can tap other trees—silver maples, black walnuts, and sycamores all produce sweet, edible sap—so if you want to experiment with making other syrups, you have several options.

↖ Sugar maples make a stunning landscape specimen. ↑ The fall color of a sugar maple. ↗ Tapping sugar maples is a time-honored tradition. You can use old-school equipment for the job, or opt for a newer setup.

LINDEN
Tilia species

Layer: Canopy tree

Edibility: Seeds, flowers, and leaves

Light requirements: Sun

Size: To 50 feet (15 m) or more

Habit and form: Deciduous tree

Flowering and fruiting times: Flowers in spring, fruits developing in early summer

Native range: Different species native around the Northern Hemisphere

Lowest hardiness temperature: −40°F/−40°C

Ecological benefit: Flowers very popular with bees and other pollinators

Pest and disease: None

Propagation: Seed

Pollination requirements: Self fertile

Lindens, also known as basswoods and lime trees, are popular street trees because they tolerate difficult conditions and have attractive, tidy growth habits. The small, greenish flowers in the spring are wonderfully fragrant and adored by bees and other pollinators. The flowers are followed by small, hard, green nutlets that will eventually turn brown. The cool thing about lindens is that their seeds can be prepared for eating. Picked when they are green, roasted until brown and fragrant, and ground, they taste like chocolate! You can grind them up and sub them for cocoa in baked goods, ice cream, or any other chocolatey dessert.

Before those chocolatey fruits appear, the linden produces flowers, which are small, white, and wonderfully fragrant. Pick them when fresh and dry them to create a floral tea.

The leaves in early spring are the best product of this tree. When they emerge and are an inch or two (2.5 to 5 cm) across, the leaves are edible and delicious. Their flavor and texture is much like a lettuce, and they can be used exactly the same way. They don't require processing to make them edible—just pick them off the tree and throw them in your salad bowl!

⬆⬆ Linden trees offer shade beneath their large canopy. ⬉ Harvested linden flowers ready for preparation into a floral tea. ⬈ Linden seeds, when roasted and ground, taste just like chocolate.

WALNUT

Juglans species

Layer: Canopy tree

Edibility: Seed

Light requirements: Sun

Size: To 50 feet (15 m) or more

Habit and form: Deciduous tree

Flowering and fruiting times: Flowers in spring, fruit ripening in fall

Native range: North America, Europe, and Asia

Lowest hardiness temperature: –40°F/–40°C

Ecological benefit: Host for several caterpillar species. Squirrels enjoy the nuts.

Pest and disease: None

Propagation: Seed

Pollination requirements: Self fertile

The English walnut (*Juglans regia*) from Europe is the most commonly eaten species, but the North American black walnut (*Juglans nigra*) is edible and tasty, as are several Asian species, including *J. cathayensis* and *J. mandshurica*. All the walnuts have beautiful foliage, though they can be a bit difficult to garden around because they release a chemical into the soil called juglone that can slow and harm the growth of other plants.

How much of an impact this actually has on plants in the garden is up for debate, and there are widely differing reports on what plants can and cannot grow well near a walnut tree. Do your research, but you'll also have to experiment to see what works well in your particular soils and climates.

Despite the irritating juglone, walnuts are tasty. You'll have to work for those nuts, especially black walnuts, which are never easy to extract from their shells. Once shelled, the nuts themselves are delicious and full of healthy oils. And freshly harvested walnuts are delicious. The oils are prone to fermentation, which is why walnuts from the store often have a bitter, off flavor to them. Harvest your own walnuts and pop them in the refrigerator or freezer right after you release them from their shells to keep them fresh and sweet tasting.

You can also try harvesting the green, unripe, walnuts to make nocino, an Italian liqueur that is dark and flavorful. I make mine from the unripe walnuts harvested from a neighbor's property.

🡔 The "meat" of walnuts is delicious. 🡕 Black walnuts are a challenge to free from their shells, but they are worth the effort.

OAK

Quercus spp.

Layer: Canopy tree

Edibility: Seeds

Light requirements: Sun

Size: To 50 feet (15 m) or more

Habit and form: Deciduous tree

Flowering and fruiting times: Flower in spring, acorns ripening in fall

Native range: Throughout the Northern Hemisphere

Lowest hardiness temperature: −40°F/−40°C

Ecological benefit: Keystone species, host plant for hundreds of butterflies and moths

Pest and disease: Sudden oak death in some areas

Propagation: Seed

Pollination requirements: Self fertile

↗ Oaks offer edible acorns for humans and wildlife alike, in addition to leaves that serve as a source of food for dozens of different insect species. ↘ Acorn flour. ↘↘ It's essential that you properly process acorns after harvesting to remove the bitter tannins.

If you're lucky to have an oak tree in your yard, celebrate it. Oaks are beautiful, enduring trees with strong wood that is rarely damaged in storms. They're also hugely important for local ecosystems. Anywhere they are native, oak trees are the primary host plant for caterpillars, which will become butterflies and moths—often literally hundreds of different species make use of a single tree.

You can enjoy your oak even more by eating the delicious, nutritious nuts it produces: acorns. These have been eaten by many cultures around the world. Acorn-flour noodles are still enjoyed in Korea, and Native Americans are thought to have informally cultivated trees with the largest, sweetest nuts.

Acorns require processing before you can eat them, as they contain a lot of tannin. The simplest way to process them is to crack the shells off with a good smack from a hammer and cover them with water. The water will turn dark as the tannins leach out. Bring the water to a boil, then pour it off, add fresh water, and keep doing that four or five times until the water stops turning dark and the nuts have a sweet, chestnut-like flavor.

The Subcanopy Layer

The subcanopy is a layer of small trees. Think redbuds (*Cercis* spp.), dogwoods (*Cornus* spp.), vine maples (*Acer circinatum*), and many common fruit trees. In nature, these species grow at the edges of the canopies of larger trees or, in the case of many fruit trees, come from dry steppe habitats where there are fewer large trees.

In the garden, you can mimic how these trees grow in nature by placing them at the edges of the shade created by your canopy trees, but subcanopy trees will also grow and thrive in full sun, often flowering and fruiting more heavily there.

Use subcanopy trees where you need beneficial shade and to gain extra height and screening from foliage, where you want a more compact package than a larger canopy tree. In small gardens, you may not have room for something larger or taller, so your subcanopy trees may be your largest layer.

Compared to the canopy layer, subcanopy trees mature faster, are easier to move and replace, and often (but not always) have shorter life spans. That being said, they're still a big investment that will live for many years. For this reason, you should take time to research what trees you want and purchase healthy specimens with visible root flares at the base of their trunks.

Unlike a canopy tree, the subcanopy trees are much less likely to do serious damage when they fall at the end of their life, so as long as a tree is still providing for you with beauty and food, you can keep them in the garden even as they show their age or the imperfections that come with improper planting or pruning.

When placing smaller, subcanopy trees, remember that you're not limited by their natural growth habit. Pruning and training can radically transform the shape and growth habit of these trees. This is much less practical with large canopy trees, but smaller trees can be managed with the help of a ladder. In the extreme, this can take the form of espalier, where branches are pruned and then tied to a structure. This will render them essentially two dimensional: a flat wall of growth. Pruning can also stop trees from growing into power lines, manage where you have shade or sun, or keep branches laden with fruit from hanging over seating areas and walkways.

Redbud trees are a great addition to the subcanopy layer. Their beautiful spring blooms are edible and support early pollinators.

The options are many here and are a great way to pack more of this layer into a small space. The only caveat is, once you start pruning, you need to commit to keeping it up. An espalier looks beautiful if pruned regularly, but missing just a couple pruning sessions will lead to its transformation into a rangy disaster. Take on new pruning projects slowly and be sure that what you do now doesn't lead to more work in the future.

If left to their own devices, many fruit trees will grow large, making it difficult to get up into them to harvest. With this in mind, take stock of how high you can easily reach from your favorite ladder, and prune growth to below that point. There's no point to maintaining a tree full of delicious fruit if you can't access half of it until it falls half rotten to the ground. Or, worse, until the fruit becomes a feast for animals like raccoons, rats, or (in my case) bears, as they can easily reach those high spots in the tree.

The small apple trees in this mixed garden are part of the subcanopy tree layer.

Many common fruit trees, such as apples and pears, can also be purchased as grafts onto dwarfing rootstocks. These are created by propagators who take branches of a fruit variety bred to be large and delicious—called a scion—and then physically attach it to the lower portion of another variety, called the rootstock. Generally, rootstocks are chosen not for their fruit, but for their health, vigor, and compact size. Growing roots actually produce plant hormones that travel up through the plant and tell the branches to grow, so a compact rootstock can miniaturize the branches of any variety grafted onto it.

Choosing a plant grafted onto a rootstock that will grow to the size that you want will limit the amount of pruning you have to do and keep the plant fitting into the layer intended for it. For example, in my garden I have a Cox's Orange Pippin apple grafted on a dwarf M26 rootstock that has height range of 8 to 12 feet (2.4 to 3.6 m) at maturity. My Melrose apple is grafted on a super dwarf M27, which tops out at 6 feet (1.8 m), ideal for a small garden and possibly a large container.

The common medlar (*Mespilus germanica*) is another subcanopy tree that produces edible fruits. I grow it in my garden, but it is not very common.

BLACK MULBERRY
Morus nigra

Layer: Subcanopy

Edibility: Fruit and leaves

Light requirements: Sun

Size: 30 feet (9 m) tall and wide (dwarf forms 6 feet/2 meters)

Habit and form: Deciduous tree

Flowering and fruiting times: Flowers in early spring, fruit throughout summer

Native range: Southwest Asia and Mediterranean Europe

Lowest hardiness temperature: −20°F/−29°C

Ecological benefit: Birds feed on the fruit

Pest and disease: None serious

Propagation: Cuttings or grafting of named forms

Pollination requirements: Self fertile

↗ A harvest of white mulberries from my neighborhood. ↗↗ Ripe mulberries are sweet and delicious. ↘ Mulberries are often produced so prolifically on the trees that there is plenty to share with the birds.

Mulberries get a bad reputation because of the white mulberry (*Morus alba*), which is a fast-growing, invasive weed. It seeds itself around vacant lots, staining sidewalks purple with its abundant but not very flavorful fruit.

While the white mulberry is a pest, the black mulberry is a very different tree indeed. A slower-growing tree that produces large, blackberry-like fruits over a very long period in the summer. It's resistant to pests and diseases, and is easy to grow. If the full-sized version is too big for your space, you can also try some of the various dwarf forms. They won't grow much taller than you, yet they still produce large yields of delicious berries.

In addition to their abundant fruit, the black mulberry's leaves can be harvested, dried, and used to make a delicious tea. The flavor is similar to green tea, but without any of the caffeine.

HAZELNUT

Corylus species

Layer: Subcanopy

Edibility: Seeds

Light requirements: Sun to part shade

Size: 8 to 10 feet (2.4 to 3 m) tall and wide

Habit and form: Deciduous tree, some forms suckering

Flowering and fruiting times: Flowers in winter into early spring, nuts ripen in late summer

Native range: North America, Asia, and Europe

Lowest hardiness temperature: −25°F/−32°C

Ecological benefit: Host plant for over a hundred species of butterflies and moths

Pest and disease: Eastern filbert blight

Propagation: Dig and separate suckers

Pollination requirements: Self fertile

There are several species of hazelnut. Those grown most commonly are the European *Corylus avellana* and the American *Corylus americana*. Both appear in form from tall shrubs to small trees, tending to sucker to form brushy colonies, and are great for informal hedging and screening. The fallen leaves make an excellent, easy-to-use mulch for garden beds over the winter.

Both species produce abundant and delicious hazelnuts that ripen at the end of the summer. Most hazelnuts grown commercially are the European species. You can find selections of this species with larger nuts, but, if you garden in North America, you're probably better off growing *Corylus americana*. This is because a disease called Eastern filbert blight occurs in the Americas: Lethal to most species of *Corylus*, the American species of hazelnut is immune.

Whatever species you choose, be sure to harvest the delicious nuts before your local squirrels get to them. Roast them for the iconic flavor that pairs so perfectly with coffee and chocolate.

↑↑ Hazelnut trees can be pruned to grow more like a vase-shaped shrub or to grow in tree form. ↖ The nuts are formed inside these frilly cases. They mature in the fall. ↗ A harvest of ripe hazelnuts.

PAWPAW

Asimina triloba

Layer: Subcanopy

Edibility: Fruit

Light requirements: Sun to shade

Size: 10 feet (3 m) tall and wide

Habit and form: Suckering deciduous tree

Flowering and fruiting times: Flowers in early spring, fruit in midsummer

Native range: Eastern North America

Lowest hardiness temperature: –20°F/–29°C

Ecological benefit: Host plant for the zebra swallowtail butterfly

Pest and disease: None serious

Propagation: Cuttings or grafting of named forms, or from seed

Pollination requirements: Need two or more varieties for good fruit set

Pawpaws are a rare temperate-zone species of a mostly tropical family. They give a bold, tropical look—as well as flavor—to cold climate gardens. Pawpaws grow as small understory trees in the wild, but adapt well to light shade or full sun in the garden.

The leaves of the pawpaw are large and tropical looking, turning an attractive yellow in the fall. In early spring, they bloom with flowers best described as unusual rather than beautiful: nodding brown bells that soon develop into large green fruits. When ripe, the fruits' skin color shifts a little toward yellowish green and they become soft to the touch. Cut them open and you'll find a soft, custardy, orange-yellow flesh. The flavor is tropical, often compared to banana, pineapple, and mango. It is delicious eaten fresh or used in cooking the way you would mango.

Be aware that raccoons and other mammals are voracious fans of the fruit, as much as humans, so you may have to watch your plants carefully to make sure something else doesn't make off with your harvest.

Pawpaws tend to sucker, forming a large thicket over time. If you have the space, let them spread. If not, cut off any suckers as they form to keep it to a single small tree.

← The flesh of pawpaw fruits is textured like custard with a tropical flavor.
↑ Pawpaws can grow in light shade or full sun.

PERSIMMON

Diospyros kaki and *D. virginiana*

Layer: Subcanopy

Edibility: Fruit

Light requirements: Sun to part shade

Size: To 30 feet (9 m) tall and wide

Habit and form: Deciduous tree

Flowering and fruiting times: Flowers in early spring, fruit ripens in late summer

Native range: East Asia and eastern North America

Lowest hardiness temperature: 0°F/−18°C (Asian), −25°F/−32°C (American)

Ecological benefit: Host plant for luna moths; fruit eaten by wildlife

Pest and disease: None serious

Propagation: Cuttings or grafting of named forms

Pollination requirements: Separate male and female plants required for pollination, but self-fertile selections exist as well

↗ Persimmon trees can produce a lot of fruits. ↗↗ The American persimmon tree grows quite large and is quite cold tolerant. → The bright orange fruits are just gorgeous.

Persimmons are beautiful small trees with an elegant shape that really show off in the fall when their leaves drop and leave behind brilliant orange fruit covering the tree abundantly like ornaments. In Asia, *Diospyrus kaki* has long been cultivated for its fruit, and there are numerous seedless selections that bear large fruits and can be enjoyed both when the fruit is still crisp like an apple and after it has ripened to a soft, custardy texture.

There are fewer selections of the American species, whose fruits tend to be smaller and have an extremely unpleasant, astringent taste until they are fully ripe and very soft. Once fully ripened, though, they are sweet and delicious. The American species is also far more cold tolerant than its Asian counterpart.

QUINCE
Cydonia oblonga

Layer: Subcanopy

Edibility: Fruit

Light requirements: Sun to part shade

Size: 12 to 15 feet (3.6 to 4.5 m) tall and wide

Habit and form: Deciduous tree

Flowering and fruiting times: Flowers in early spring, fruit in late summer

Native range: Western Asia/East Europe

Lowest hardiness temperature: –20°F/–29°C

Ecological benefit: Nectar source for orchard mason bees

Pest and disease: None serious

Propagation: Cuttings or grafting of named forms

Pollination requirements: Self fertile

As they bedeck themselves with white or pink flowers in springtime, you'll have no question why quince trees are a great addition to the garden. The blooms look much like a crabapple or pear, but they're much larger. The foliage has a nice, bold texture through the summer, and in late summer it produces the quince fruit, which appear like large, lumpy yellow apples.

The fruit of most varieties of quince is hard, tart, and very astringent, even when ripe; but when cooked, it becomes soft and delicious, with a strong fragrance. Add some to an apple sauce or apple pie to really enhance the flavor.

One of my favorite ways to enjoy quince is to make membrillo, a paste made from the fruit. To make this delicacy, boil the quince until tender, puree it, and cook down with sugar until thick and rose colored. Spread this in a pan and place in a warm (not hot) oven to dry out. The final product will be a firm, sliceable block that levels up any cheese board. It is traditionally paired with a firm sheep cheese like manchego.

Be aware that quince is different from the more commonly planted flowering quince (genus *Chaenomeles*), which has more abundant, brilliantly colored flowers, but fruit that's less tasty (though still edible).

↖ Quince tree coming into flower. ↑ Quince is a perfect subcanopy small tree. ↗ These fruits are almost ready to harvest.

JUJUBE
Ziziphus jujuba

Layer: Subcanopy

Edibility: Fruit (single-seeded drupe)

Light requirements: Sun

Size: 15 to 30 feet (4.6 to 9 m) tall and wide

Habit and form: Deciduous tree

Flowering and fruiting times: Flowers in early spring, fruit ripens in late summer

Native range: Southeast Europe and China

Lowest hardiness temperature: −20°F/−29°C

Ecological benefit: Dense roots prevent erosion

Pest and disease: None serious

Propagation: Cuttings or grafting of named forms

Pollination requirements: Partially self fertile, having two or more varieties improves cross-pollination for better yields

Jujubes are an attractive shade tree with glossy green leaves and abundant small round fruits about the size of dates. Popular and well established in Asia, they're less known in the rest of the world, but are still well worth growing and enjoying.

When the fruits turn a rich shade of brown they are ripe. At this stage, they have a crisp texture and a flavor somewhat like an apple. More often, however, they are allowed to dry, which gives them a flavor and texture similar to dates. This is the source of one of their common names, the Chinese date.

Dried jujubes can be used in the same way you would dates or raisins in sweet or savory dishes. In Asian cuisines they're used in desserts as well as savory soups.

↑↑ A jujube tree. ↖ Jujube fruits ready for harvest. ↗ Dried jujubes ready to enjoy.

APPLE AND CRABAPPLE

Malus spp. and hybrids

Layer: Subcanopy

Edibility: Fruit

Light requirements: Sun

Size: 15 feet (4.5 m) tall and wide (varies by variety)

Habit and form: Deciduous tree

Flowering and fruiting times: Flowers in early spring, fruit ripen in late summer

Native range: Central Europe

Lowest hardiness temperature: –30°F/–34°C or lower

Ecological benefit: Nectar source for early pollinators; fruits enjoyed by wildlife

Pest and disease: Rust, mildew, and other fungal diseases.

Propagation: Grafting of named forms

Pollination requirements: Two or more varieties required for pollination and fruit set

Apples need no introduction as a fruit, and they make a perfect addition to an edible landscape. Crabapples and regular apples both produce edible fruit and beautiful spring floral displays. Your choice depends on the focus of your landscape.

Crabapples are tops for floral display, but the fruit they produce is small, though still edible and great as the basis for crabapple jelly. Larger apples will not match crabapples for floral impact, but the larger fruit is easier to harvest and generally tastier.

A showy crabapple may be your best choice for a highly visible front garden, while more practical apples might work better in a more utilitarian space behind the house.

You also have a huge selection when it comes to the size and shape of your apple trees. Apples are grafted onto rootstocks, which will determine their final size and growth habit. This means you can choose from large trees to medium-size ones, all the way down to dwarfing rootstocks that will top out at 12 feet (3.6 m). In small spaces, you can also prune them to grow flat against a fence or wall using the espalier method. This can transform a fruit tree into a beautiful living sculpture and also provide you with delicious fruit in a small space.

The same is true for pears (*Pyrus prunus*), cherries, plums, apricots, and peaches (*Prunus* spp.). All these traditional fruit trees have showy spring flowers, are available in dwarf versions, and will fit perfectly in a beautiful layered edible garden.

←← Apple trees are a lovely addition to an edible landscape. ← Spring flowers on a small apple tree. ↑ Ready for harvest.

DOGWOOD

Cornus spp.

Layer: Subcanopy

Edibility: Berry-like fruit

Light requirements: Sun to medium shade

Size: 15 feet (3.6 m) tall and wide

Habit and form: Deciduous tree

Flowering and fruiting times: Flowers in early spring, fruit ripening in midsummer

Native range: Species native around the Northern Hemisphere

Lowest hardiness temperature: −30°F/−34°C (*Cornus mas*)

Ecological benefit: Spring nectar source, fruit enjoyed by birds

Pest and disease: Dogwood anthracnose effects some species

Propagation: Seed or grafting of named forms

Pollination requirements: Self fertile

↗ *Cornus mas* in bloom. ↘ The fruits of *Cornus mas* are ripe and ready. ↘↘ An ample harvest.

Dogwoods are famous for their spring flowers, but they also produce edible fruits. The North American species (*Cornus florida* in the East and *Cornus nuttallii* in the West) have blooms with showy white or pink bracts, and edible fruit (though it's not all that tasty). *Cornus kousa*, the popular Asian dogwood, has large, rounded fruits with a tough bitter skin surrounding tasty, sweet flesh.

The real winner for edibility in this species is the European (*Cornus mas*). Lacking large, showy bracts, this shrub or small tree has clusters of small yellow flowers in late winter and early spring. These give way to bright red fruits in summer, sometimes called Cornelian cherries. The flavor is pleasant but extremely strong and very sour. You may not like eating them straight off the tree, but juiced and sweetened or added to baked goods they add a wonderful shot of flavor.

MAGNOLIA
Magnolia spp. and hybrids

Layer: Subcanopy

Edibility: Flowers

Light requirements: Sun to part shade

Size: 15 feet (3.6 m) tall and wide (varies by species and cultivar)

Habit and form: Deciduous and evergreen trees

Flowering and fruiting times: Flowers in early spring and summer

Native range: Asia and the Americas

Lowest hardiness temperature: −25°F/−32°C with some variation depending on species

Ecological benefit: Seeds eaten by squirrels and wild turkey

Pest and disease: None serious

Propagation: Cuttings or grafting of named forms

Pollination requirements: NA

Magnolias are some of the most beautiful small trees you can have in your landscape. The popular deciduous species and hybrids (mostly *M. kobus* and *M. × soulangeana* and *M. sprengeri*) cover themselves with huge pink or white blooms in the spring. The ever-popular *Magnolia grandiflora* is evergreen, featuring enormous, fragrant flowers in summer.

The flower petals are edible, though their exact flavor will vary depending on the specific variety. In general, they combine lemon, spicy ginger or cardamom, and floral flavors. White-flowered varieties tend to have a milder flavor than the darker-colored blooms. They can be tossed into a salad, or simmered in sugar and water to make a spicy simple syrup to add to drinks or baked goods.

↗ The pink and flavorful petals of *M. × soulangeana*. → The edible petals of the Southern magnolia, *Magnolia grandiflora*. →→ A springtime magnolia petal harvest ready to turn into a sweet syrup.

The Shrub Layer

The shrub layer is one of the most versatile and hard-working layers in your garden. They offer beauty and edible features, while also requiring some of the lowest maintenance for most gardens. Their height—especially with combined with a thick layer of mulch—means that shrubs, when established, can outcompete and out-shade virtually all weeds. Shrubs' medium size also makes them easy to work around, so that you won't have to get down on your knees or drag out a ladder to prune or harvest them.

The range of useful and beautiful shrubs you can include in your garden is huge, and many will provide attractive foliage and beautiful flowers, along with their edible fruits. Evergreen shrubs, in particular, provide a great year-round structure and beauty in a garden.

There are many types of edible shrubs, including common ones like blueberries (*Vaccinium corymbosum*) and currants (*Ribes* spp.) and more unusual choices such as this sea buckthorn (*Hippophae rhamnoides*).

Hedgerows can be planted to delineate property lines or screen off neighbors. Why not use edible shrubs or trees in your hedgerow instead of standard evergreen or flowering shrubs? Here, pear trees along a fence line shelter cut flowers and edible greens from the wind.

Foundation Plantings

Shrubs are also a classic ingredient of foundation plantings, the traditional skirt of plants around the base of a house.

Boxwoods (*Buxus* spp.), though not edible, are a traditional foundation planting shrub. Another option are yews (*Taxus baccata*), which are toxic but can be mixed with other, edible shrubs that will serve the same visual effect. The humble juniper (*Juniperus* spp.) is a great example, a wonderful evergreen shrub that will fit in a traditional foundation planting, but bring with them brightly flavored "berries" (technically they are cones) to can after harvesting. Tea plants (*Camellia sinensis*) make beautiful evergreen foundation hedges that can be pruned and harvested at the same time.

Hedgerows

Laying out your shrubs as hedgerows is one of the most useful ways to place these plants. A hedge can be formal—a row of identical plants growing tightly into a living wall—but they can also be configured in a more naturalistic pattern. Mix and match different species and varieties and, instead of shearing them tightly, let them grow to their natural form or lightly prune them to control size and encourage branching.

Whatever form a hedgerow takes, this format offers a number of advantages. Privacy is the obvious one, screening out noisy or nosy neighbors and blocking less-than-attractive views. Hedges are also great noise barriers, dampening the sounds of busy streets or neighbors as well.

Hedges can be a wonderful windbreak, too, actually slowing down and limiting winds better than a solid wall or fence. In you live in an open, windy area, this can make enjoying your garden more comfortable, and some plants will also appreciate a good windbreak. Broad-leaved evergreens will survive cold winters much better if sheltered from drying winter winds.

The dense branching structure of a hedge is also prime habitat for wildlife, particularly songbirds. They usually prefer spaces like this to nest and roost, since they offer protection from larger predators like hawks. Include some shrubs that produce berries in your hedgerow and birds will find a welcoming place to enjoy your garden.

In this garden, edible plants include a plum tree (*Prunus domestica*), hollyhocks (*Alcea rosea*), roses (*Rosa* spp.), Shasta daisy (*Leucanthemum* x *superbum*), whose greens are quite tasty cooked, and squash (*Cucurbita* spp.).

STRAWBERRY TREE

Arbutus unedo

Layer: Shrub

Edibility: Berries

Light requirements: Sun

Size: 6 to 15 feet (1.8 to 4.6 m) tall, 6 to 8 feet (1.8 to 2.4 m) wide

Habit and form: Upright evergreen shrub

Flowering and fruiting times: Flowers in fall through winter with fruit developing throughout the flowering period, ripening the following year

Native range: Europe and Mediterranean regions

Lowest hardiness temperature: 0°F/−18°C

Ecological benefit: Flowers for pollinators

Pest and disease: None serious

Propagation: Cuttings

Pollination requirements: Self fertile

↗ The deep red fruits are ready for harvest. → A strawberry tree in bloom. →→ A handful of soft fruits ready to turn into jam.

This broadleaf evergreen shrub is often overlooked and underappreciated when used in mass commercial plantings. But when planted as a small specimen shrub or tree in a garden setting, it's easy to appreciate its highly ornamental, four-season qualities: white clusters of pendulous, urn-shaped flowers (similar to blueberries) that support hummingbirds over the fall–winter months; round textured fruit that develops during the flowering period; and peely reddish-brown bark as well as dark green glossy leaves throughout the year.

Strawberry trees are easy to maintain and can tolerate a wide range of conditions. They do best when provided with full sun and well-draining spot, conditions similar to the warm Mediterranean regions from which it is native. Fruit from the previous year's flowers will be ready for harvest when they soften and turn a vibrant orange-red, resembling small spiky lychee (*Litchi chinensis*) fruit. The edible fruit is mild in flavor with a slightly mealy texture, but it can be transformed into delicious jams, syrups, and fermented drinks.

TEA
Camellia sinensis

Layer: Shrub

Edibility: Leaves

Light requirements: Sun to shade

Size: 4 to 5 feet (1.2 to 1.5 m) tall and wide

Habit and form: Evergreen shrub

Flowering and fruiting times: Flowers in fall, winter; white flowers are fragrant

Native range: East Asia

Lowest hardiness temperature: 0°F/−18°C

Ecological benefit: Bushes sequester 50% of the atmospheric CO_2 in their biomass

Pest and disease: Prone to various fungal diseases and leaf spots

Propagation: Cuttings

Pollination requirements: NA

↗ A young, recently planted tea bush. ↗↗ The simple white flowers of *Camellia sinensis*. ↘ The new growth is plucked off and used for making homegrown tea.

Tea makes a fantastic landscape shrub, with glossy evergreen leaves and attractive (though not overwhelming) white flowers in the fall.

One of the best features of this plant as part of an edible landscape is the way tea is harvested. Only the newest growth—just the bud and one or two leaves below it—is picked while they are still tender and delicate. These are harvested multiple times throughout the year. That regular picking off of new growth is also how you shear a shrub into a tight, formal hedge, so this is a perfect choice if you want an edible landscape but also want an orderly, pruned look in the garden. Your neighbors will just think you're keeping up on the pruning when you're actually harvesting delicious homegrown tea.

Once picked, you can crush the leaves and let them oxidize for twenty-four to forty-eight hours, then dry them to make black tea, or pop them briefly in a steamer basket and dry them to make green tea. Tea grown in the shade or harvested during cool times of the year makes the most delicately flavored green tea, while plants grown in full sun or picked during hotter months makes a more robust black tea.

GOJI, AKA GOJI BERRY

Lycium barbarum

Layer: Shrub

Edibility: Berries and leaves

Light requirements: Sun

Size: 3 to 6 feet (1 to 2 m) tall and wide

Habit and form: Loose, open shrub

Flowering and fruiting times: Purple flowers in spring followed by berries

Native range: East Asia

Lowest hardiness temperature: –15°F/–26°C; some varieties are even hardier

Ecological benefit: Bees and hummingbirds visit for the nectar

Pest and disease: Can get powdery mildew if situated in a location with poor air circulation or with temperature and humidity fluctuations; not generally bothered by insect pests

Propagation: Cuttings or seed

Pollination requirements: Self fertile

↗ Goji flowers and berries are very attractive. ➔ The arching branches of goji shrubs. ➔➔ A harvest of gojis.

Goji berry is often promoted as a health food for its tart, nutrition-packed red berries. Easy to grow, durable, and tolerant in a wide range of climates, the goji berry should be planted in an area with less rich soil—the richer the soil, the more likely you'll encourage rank, lush growth over fruit production.

When shopping for goji berry shrubs, look for named cultivars that have been selected for good, tasty fruit production. Sometimes goji berry is grown from seed, but seedling forms may produce minimal fruit that is far less tasty than the yield from named varieties. If you harvest more than you can eat, the berries can be dried or frozen to preserve them.

The leaves of goji berry are edible as well. They are traditionally cooked in soups in China, often combined with dried goji berries, ginger, stock, and chicken or other meat. Just watch out for the thorns while harvesting!

HARDY FUCHSIA

Fuchsia magellanica

Layer: Shrub

Edibility: Flowers and berries

Light requirements: Shade

Size: 3 to 4 feet (1 to 1.2 m) tall and wide

Habit and form: Evergreen shrub

Flowering and fruiting times: Flowers all summer, followed by berries

Native range: Southern tip of South America

Lowest hardiness temperature: −10°F/−23°C

Ecological benefit: Provides nectar for hummingbirds

Pest and disease: Aphids and whiteflies can be a problem

Propagation: Cuttings

Pollination requirements: NA

↗ The graceful arching stems of hardy fuchsia. ↘ A hardy fuchsia fruit. ↘↘ The blooms are hummingbird magnets.

Hardy fuchsia is a terrific plant for the edible landscape thanks to its stunning beauty, decorated with dangling flowers that are hummingbird magnets throughout the summer. After they fall off, the plant is covered with attractive red berries. Even better, it thrives in shaded conditions where it can often be hard to grow much food.

Both the berries and the flowers are edible and tasty. The flowers have a peppery flavor and make a beautiful addition to a salad. The fruits are sweet and tasty. In addition to the hardiest species, *F. magellanica*, there are numerous other species and hybrids, all of which have edible berries, though some are tastier than others.

Hardy fuchsia is a fantastic edible plant where it thrives. It requires winters where temperatures don't regularly drop below 10°F (−12°C) and prefers cool summers where temperatures don't get above 86°F (30°C). In colder climates, you can grow fuchsias in containers and overwinter them indoors or just treat them as annuals.

GOUMI
Elaeagnus multiflora

Layer: Shrub

Edibility: Berry-like fruit (a drupe, not berry)

Light requirements: Sun

Size: 6 to 10 feet (1.8 to 3 m) tall and wide

Habit and form: Deciduous shrub

Flowering and fruiting times: Fragrant flowers in early spring followed by summer berry-like fruit

Native range: East Asia

Lowest hardiness temperature: −20°F/−29°C

Ecological benefit: Fruits enjoyed by birds and mammals

Pest and disease: None serious

Propagation: Cuttings

Pollination requirements: Partially self fertile, cross pollination with compatible variety planted in close proximity helps with yields

This shrub is a fantastic choice for difficult, low-fertility sites. Not only is goumi extremely drought tolerant once established, it also actinorhizal, meaning that its roots can form a symbiotic relationship with a little organism called *Frankia* (an actinomycetota) in the soil. Similar to legumes like peas and beans, this relationship allows the *Frankia* to transform gaseous nitrogen in the air into a form that the goumi can use. In return, the goumi provides a place for the *Frankia* to live. The takeaway for you as a gardener is that goumi can, essentially, produce its own nitrogen fertilizer.

It is an attractive plant with dense leaves dusted with silver, most prominently on the underside, which are a stunning, shimmery silver. The ivory-gold flowers in spring aren't large, but they are very fragrant, filling the garden with the scent of orange blossoms and jasmine. After the flowers comes the sweet, tart, delicious berry-like fruit (drupes). Be sure to wait until those fruits are soft and fully ripe, as picked too soon they'll be astringent and unpleasant. Once ripe, though, they are sweet and tart, great fresh off the plant and with a flavor strong enough to work well cooked in jam or pies.

I really love this plant, both for its delicious fruits and just how beautiful it is. Every part of it has a shimmery metallic appearance. The stems are textured bronze, the leaves silver (they make the most beautiful mulch when they drop to the ground!), and even the fruits have gold flecks on them.

← Unopened goumi flowers. ↖ Goumi fruits nearly ready for harvest.
↗ A beautiful harvest from a beautiful shrub.

SERVICEBERRY

Amelanchier alnifolia, and other species

Layer: Shrub

Edibility: Berry-like fruit (pomes)

Light requirements: Sun to light shade

Size: To 6 feet (1.2 to 2 m) tall and wide

Habit and form: Deciduous tree

Flowering and fruiting times: Flowers in early spring, followed in early summer by small, berry-like fruit

Native range: Western North America

Lowest hardiness temperature: −30°F/−34°C

Ecological benefit: Fruits are favorites of birds

Pest and disease: May be prone to leaf spot, fire blight, and powdery mildew in some climates

Propagation: Cuttings or seed

Pollination requirements: Self fertile

Various species of serviceberry are native across different parts of North America, and they are all fantastic large shrubs or small trees.

In spring, they bloom with clouds of white flowers that give way to clusters of purple or red berry-like fruit, technically called pomes. They look and taste a lot like blueberries. But unlike blueberries, they don't demand acidic soil to thrive.

In fall the leaves turn shades of yellow, orange, and red for a fantastic display. The bark is smooth silver, making every part of the plant an attractive garden addition.

Many service berries will grow into large trees, but you can look for more compact forms at the nursery. *Amelanchier alnifolia* 'Regent' is an excellent small form.

← A sweet harvest. ↖ Serviceberries are a personal favorite. ↗ Serviceberries produce lovely white spring flowers.

BLUEBERRY

Vaccinium spp. and hybrids

Layer: Shrub

Edibility: Berries

Light requirements: Sun

Size: 3 to 4 feet (1 to 1.2 m) tall and wide

Habit and form: Deciduous shrub

Flowering and fruiting times: Flowers in early spring followed by berries in the summer

Native range: Different species native to much of the Northern Hemisphere

Lowest hardiness temperature: −30°F/−34°C

Ecological benefit: Early spring nectar source for bees

Pest and disease: Leaves yellow in alkaline soils

Propagation: Cuttings

Pollination requirements: Self fertile, but set fruit better with multiple varieties grown together

I don't need to sell you on blueberries as an edible crop. They're delicious, packed with nutrients, and equally wonderful eaten by the handful or added to a muffin or other baked good.

What you might not realize is that blueberries are fantastically ornamental as well. Their growth habit is a tidy, attractive shrub, with shiny green foliage. The white flowers in spring are attractive, as are the berries of course, and in the fall many blueberries will put on a foliage show that rivals the best burning bush or maple. Once the leaves drop, the branches are revealed: These are a bright red color, providing a lot of beauty and interest through the winter months.

Blueberries prefer fertile soil that doesn't dry out too much, and they demand acidic soil. If your soil is naturally alkaline, you can still grow them, just choose one of the smaller varieties and grow it in a container.

Traditionally, blueberries have been the domain of northern climates, with most species languishing in areas with hot summers. But modern breeding has combined the heat tolerance of species native to southeastern North America with the tasty fruit of northern species to ensure there is a blueberry that will thrive in nearly every climate.

↑ Try low-bush or wild blueberries with cranberries if you have a low-lying area in your garden. ↗ Growing more than one variety is essential for good fruit set. → Blueberries make great partners with a ground cover of strawberries.

SEA BUCKTHORN
Hippophae rhamoides

Layer: Shrub

Edibility: Berries

Light requirements: Sun

Size: 8 to 12 feet (3 to 4 meters) tall and wide

Habit and form: Suckering shrub

Flowering and fruiting times: Flowers early spring, fruits in fall into winter

Native range: Europe and northern Asia

Lowest hardiness temperature: −40°F/−40°C

Ecological benefit: Great to stabilize soils and prevent erosion on difficult sites

Pest and disease: None serious

Propagation: Cuttings

Pollination requirements: Needs male and female plants together to produce fruit

↗ A plant that's ready for harvest.
→ The foliage of sea buckthorn.
→→ Unharvested buckthorns will cling to the tree through winter, making a beautiful display.

Sea buckthorn is a durable plant that thrives in lean soils, coastal areas. The narrow, silvery leaves are attractive. After the not-very-noticeable flowers in the spring, female plants produce big clusters of beautiful orange-yellow berries. The berries ripen in the fall and can last well into the winter for a prolonged show.

The berries are extremely high in nutrients and have a strong, tart, citrusy flavor. They're really too tart to eat fresh from the plant, but with some added sweetener the juice is delicious on its own or used in cooking.

ELDERBERRY

Sambucus spp.

Layer: Shrub

Edibility: Flowers and berries

Light requirements: Sun to light shade

Size: 4 to 6 feet (1.2 to 2 m) tall and wide

Habit and form: Suckering deciduous shrub

Flowering and fruiting times: Flowers in spring followed by summer berries

Native range: North America, Europe, and Asia

Lowest hardiness temperature: −30°F/−34°C

Ecological benefit: Nectar source for many pollinators and beneficial insects

Pest and disease: None serious, occasional mildew

Propagation: Cuttings

Pollination requirements: Self fertile

Various species of elderberry can be found throughout the Northern Hemisphere, often in moist spots. They tend to grow as suckering masses, with individual stems only living a few years before being replaced by new growth. In the spring, they produce big clouds of fragrant white flowers, which are followed by numerous tiny black berries.

Even more ornamental forms have been selected with finely cut, fernlike foliage in shades of dark purple or bright yellow-green. Though the colored foliage forms are quite beautiful, they usually produce far less fruit than the typical green-leaved species.

The flowers and fruit of the elderberry have long been used for food. The big flower clusters have a delicious aroma. One traditional way to put that to use is to make a simple syrup and steep the flowers in it for twenty-four to forty-eight hours, until the rich, floral aroma and flavor has infused the syrup.

The berries are also tasty, but slightly toxic when raw, so be sure to cook them into a pie or preserves before enjoying them.

⬆ Harvesting an elderberry cluster. ⬈⬈ The small, dark berries of the elderberry bush. ⬉ White elderflowers can be harvested and used to make a sweet syrup.

CHILEAN GUAVA
Ugni molinae

Layer: Shrub

Edibility: Berry

Light requirements: Sun

Size: 3 to 5 feet (0.9 to 1.5 m) tall, 2 to 3 feet (0.6 to 1 m) wide

Habit and form: Deciduous shrub

Flowering and fruiting times: Flowers in spring followed by early fall berries

Native range: South America

Lowest hardiness temperature: 10°F/−12°C

Ecological benefit: Flowers for bees

Pest and disease: None serious

Propagation: Cuttings

Pollination requirements: Self fertile

The small, dark green leaves of this evergreen shrub make it a handsome addition to the shrub layer in-ground or when grown in a large container. A planting of Chilean guavas makes a beautiful informal hedge that is easily pruned to fit the space.

The plant does best in sheltered areas that get full to part sun as well as winter protection in colder climates. It thrives in moist, well-draining soils but is also drought tolerant once established.

Dainty, pendulous pink-white flowers that appear in late spring are visited by bees, followed by small red berries that are ready to harvest in early fall. The berries have a very unique flavor that is commonly described as minty, strawberry bubblegum, or cotton candy—they're delicious and can be eaten raw or cooked.

↑ My Chilean guava harvest. ↗ The branches are laden with fruit.

POMEGRANATE

Punica granatum

Layer: Shrub

Edibility: Fruit

Light requirements: Sun

Size: 6 to 8 feet (1.8 to 2.4 m) tall and wide

Habit and form: Deciduous shrub

Flowering and fruiting times: Flowers in early summer followed by fruits

Native range: Europe and Asia

Lowest hardiness temperature: 0°F/−18°C

Ecological benefit: Nectar for bees and hummingbirds; habitat for small mammals

Pest and disease: Prone to fungal diseases in wet areas

Propagation: Cuttings

Pollination requirements: Self fertile

The beautiful, red jewel-like seeds of the pomegranate are a familiar food, packed with a wonderful intense flavor. Pomegranates are also attractive shrubs, with showy, brilliant orange-red flowers in the spring and into summer. The fruits themselves look like huge red Christmas ornaments growing on the plant.

Knowing when to harvest pomegranate fruits can be a challenge. The fruit won't continue to ripen once it is off the tree, so wait until it is fully ripe to get the best flavor. If you see the outside of the fruit crack, that is a good sign it is ripe and ready to harvest, though it may already be overripe at this stage and will not store as well as ones picked before they crack.

The other clues for ripeness are a little hard to follow, but easier to judge with experience. The easiest way is weight: As the fruit ripens, it fills with delicious juice. The ones ready to harvest will feel dramatically heavier than unripe ones in your hand. The skin of the fruit will also develop a softer, almost leathery texture as it matures. Your first harvest, it can be a bit difficult to judge, but you'll get the hang of it.

Pomegranates are drought tolerant and don't do well in soggy soils. Otherwise they're not picky. Many of the traditional varieties aren't cold tolerant, but there are cold-hardy selections originating in eastern Europe and Russia that greatly expand the range of where this beautiful and delicious plant can be grown. I'm currently growing the variety 'Salavatski', which is supposed to be one of the hardiest.

In cold climates, it can be a challenge to get the fruits to ripen before the end of a short growing season, so try positioning them against a warm, south-facing wall to give them the most heat possible during the summer months.

← Sweet little flowers. ↑ Developing pomegranates.

BAY LAUREL
Laurus nobilis

Layer: Shrub

Edibility: Leaves

Light requirements: Sun to shade

Size: To 30 feet (10 m) tall, but kept smaller by pruning

Habit and form: Evergreen shrub or tree

Flowering and fruiting times: Not significant

Native range: Mediterranean Europe

Lowest hardiness temperature: 0°F/–18°C

Ecological benefit: Cover and nesting sites for birds

Pest and disease: Susceptible to phytophthora root rots and leaf spot when planted in overly wet, poorly drained conditions

Propagation: Cuttings

Pollination requirements: NA

Bay laurels are the source of the bay leaves in your spice rack, the aromatic leaves that add dimension to so many dishes. It's also a beautiful plant, with dark green, glossy evergreen leaves. The plant responds well to pruning, so you can easily keep a bay laurel pruned into a hedge or any other shape you desire.

If your winter temperatures drop much below freezing, the bay laurel will often freeze to the ground, only to come back from its roots in the spring. In colder climates, where temperatures regularly drop below 15°F (–9°C), it is best grown in a container that can be moved inside to overwinter.

↑ Bay is easy to grow in a large pot that can be moved indoors for the winter.
← A small bay tree in its first year of growth.

PINEAPPLE GUAVA

Feijoa sellowiana

Layer: Shrub

Edibility: Flowers and fruits

Light requirements: Sun

Size: 6 to 10 feet (1.8 to 3 m) tall and wide

Habit and form: Evergreen shrub

Flowering and fruiting times: Flowers in early summer, fruit in late summer or fall

Native range: South America

Lowest hardiness temperature: 5°F/–15°C

Ecological benefit: Nectar source for pollinators

Pest and disease: Prone to root rots in cool, wet climates

Propagation: Cuttings or seed

Pollination requirements: Seedlings require multiple plants for pollination, though self-fertile varieties also exist

Pineapple guava is a durable, easy-to-grow plant in warm climates. It's drought tolerant and generally pest and disease free, though it loves hot weather and can struggle in cool summer climates. The evergreen leaves have a beautiful silvery hue to them, and in early summer they bloom with extremely showy flowers with thick, fleshy, white-to-pink petals that surround a fireworks display of bright red stamens. These flowers then develop into fragrant green-skinned fruits.

The name pineapple guava comes from the fruits, which, when ripe, have a delicious tropical pineapple flavor with a hint of mint mixed in for good measure. In addition to the fruits, the flowers are edible.

For most plants, edible flowers have a mild, crisp taste, and are most often enjoyed as a salad garnish. Pineapple guava flowers are a whole other matter, and the fleshy petals are sweet—sometimes compared to marshmallows—offering a delicious treat eaten straight off the plant. If you pick the petals off and leave the rest behind, the flowers will still develop into fruit later in the season.

← Developing fruits. ↑ The white undersides of pineapple guava leaves.
↗ The pink and red blooms of the pineapple guava.

BLACK CHOKEBERRY

Aronia melanocarpa

Layer: Shrub

Edibility: Berry-like pome fruits

Light requirements: Sun to part shade

Size: 3 to 6 feet (1 to 2 m) tall and wide

Habit and form: Deciduous shrub

Flowering and fruiting times: Flowers in spring followed by berry-like fruit in early summer

Native range: North America

Lowest hardiness temperature: −40°F/−40°C

Ecological benefit: Birds love the berries and flowers feed a wide range of pollinators

Pest and disease: None

Propagation: Cuttings

Pollination requirements: Self fertile

Chokeberries are native to a wide swath of North America, thriving in a wide range of conditions from low wet areas to dry ones. Very adaptable in the garden, they form a somewhat suckering shrub with clouds of attractive white flowers in spring followed by big clusters of blackberry-like fruit. In fall, the leaves turn intense shades of red and purple, giving them interest in nearly every season of the year.

If you just pick a fruit and pop it in your mouth, you might be unimpressed. The name chokeberry refers to the fruit's intensely tart, slightly astringent flavor, especially when harvested before a frost. If you wait until after the first time temperatures dip below freezing in the fall, they'll become a little sweeter and milder in flavor.

That strong flavor is actually their great attribute. Few of us would like to eat a cranberry raw, but many love them cooked in a variety of ways. And aronia actually tastes a great deal like cranberries (*Vaccinium macrocarpon*), and it can be used in the same way. In fact, check the label of containers of cranberry juice at the grocery store—you'll find that many of them contain a fair amount of aronia juice!

Try an aronia relish instead of cranberry this fall and you'll be delighted with the result. While most of us don't have bogs in our backyards to grow cranberries, aronia will thrive in most gardens and can provide an abundant harvest of these flavor-packed fruits.

← Aronia flowers. ↑ Ripe fruits ready for harvest. ↗ A side view prior to picking the fruit.

CURRANT AND GOOSEBERRY

Ribes spp. and hybrids

Layer: Shrub

Edibility: Berries

Light requirements: Sun to part shade

Size: 3 to 6 feet (1 to 2 meters) tall and wide

Habit and form: Thorny deciduous shrub

Flowering and fruiting times: Flowers in spring followed by early summer berries

Native range: Different species around North America, Europe, and Asia

Lowest hardiness temperature: −25°F/−32°C

Ecological benefit: Nectar source for hummingbirds and butterflies; food and nesting sites for birds and small mammals

Pest and disease: Currant maggots can be a problem on the fruit, and currant sawfly larvae will eat the foliage

Propagation: Cuttings

Pollination requirements: Most self fertile, but better fruit set when multiple varieties grown together

There are many species of the genus *Ribes*, including currants, gooseberries, jostaberries, and many others. All are delicious, attractive additions to the landscape.

Whether they're black, red, or white, currants have sweet, tart berries that can be enjoyed fresh, juiced, or dried and used in baked goods. Gooseberries are larger and juicier, a bit like a grape but with a more intense flavor. The berries are beautiful as well, with currants growing in big clusters that glow like jewels.

Whichever berry plant you choose—or if you choose multiples—they grow into a thick, often thorny, shrub, with generally underwhelming flowers. These shrubs are remarkably adaptable to a wide range of climates and soil conditions. Even better, they'll still flower and fruit well even in partially shaded sites, which are often difficult places to grow most fruits.

If you garden in the eastern part of North America, you might be restricted from growing currants because they can host a disease called white pine blister rust. The disease doesn't do significant damage to the currants, but it can move from them to damage pine trees. Check with your local state extension to see if it's okay to grow currants in your area.

← Currants are an easy-to-grow choice for the edible landscape. ↑ Purple gooseberries. ↗ Currant bushes have a beautiful form.

HONEYBERRY OR HASKAP

Lonicera caerulea

Layer: Shrub

Edibility: Berries

Light requirements: Sun to light shade (shade preferable in hot climates)

Size: 3 to 6 feet (1 to 2 m) tall and wide

Habit and form: Deciduous shrub

Flowering and fruiting times: Flowers in spring, fruit in early summer

Native range: Northern North America, Europe, and Asia

Lowest hardiness temperature: −40°F/−40°C

Ecological benefit: Bumble bees forage on the blooms

Pest and disease: None

Propagation: Cuttings

Pollination requirements: Two different plants of compatible varieties required for pollination

↗ Haskaps produce small yellow flowers very early in the spring.
↗↗ You'll need at least two different varieties for cross-pollination.
↘ Haskaps ready for harvest.

The berries of this shrub look like oblong blueberries. They taste like blueberries, too, but with a little more tartness. They're delicious eaten straight off the plant or cooked as you would a blueberry in pies or other baked goods. One of the great virtues of having a honeyberry in the edible landscape is that the fruit ripens early, at the beginning of the summer, when most other fruits are still months away from being ready to harvest.

Despite offering blueberry lookalike fruits, this plant is actually a species of honeysuckle, which explains their easy-growing ways in the garden. If you don't have the acidic soil that blueberries require, definitely consider trying a honeyberry instead.

ROSE
Rosa rugosa and other species

Layer: Shrub

Edibility: Fruit, petals

Light requirements: Sun

Size: 3 to 4 feet (1 to 1.2 m) tall and wide

Habit and form: Suckering thorny deciduous shrub

Flowering and fruiting times: Flowers all summer followed by fruit

Native range: Coastal east Asia

Lowest hardiness temperature: –40°F/–40°C

Ecological benefit: Stabilizes soils against erosion

Pest and disease: None

Propagation: Cuttings, seed, division of suckers

Pollination requirements: Self fertile

↗ Rose petals are edible.
↗↗ Rose hips ready for harvest.
↘ Hips from *Rosa rugosa*.

All roses can produce edible fruits, called rose hips, after their flowers fade. Most garden roses have been bred for showy flowers, though, and the hips they produce—if they make any at all—are small, dry, and not very palatable. The tastiest rose hips are those produced by *Rosa rugosa*. This wild rose is also called a beach rose because it thrives in lean, sandy soils and is tolerant of salt spray. It has large pink or white flowers that bloom all summer and dark green crinkled foliage that colors up in the fall.

After the flowers come the rose hips: big, round, bright red fruits. Don't eat them straight off the plant, as they're a bit dry and have irritating hairs inside them, around the seeds. The classic way to enjoy them is to make rose hip tea, which infuses your hot water with a delicious, almost floral, flavor, and an impressively large dose of vitamin C.

ROSEMARY
Salvia rosmarinus

Layer: Shrub

Edibility: Leaves, flowers

Light requirements: Sun

Size: 3 to 4 feet (1 to 1.2 m) tall and wide

Habit and form: Evergreen shrub

Flowering and fruiting times: Blue flowers in later winter or early spring

Native range: Mediterranean Europe

Lowest hardiness temperature: –10°F/–23°C; 'Arp' variety is cold hardy

Ecological benefit: Flowers early food source for pollinators

Pest and disease: Root rot in wet soils

Propagation: Cuttings, seed

Pollination requirements: NA

Rosemary as an herb is widely used in cooking, and it's a fantastic landscape plant as well. The fragrant foliage is wonderful placed next to paths where you can get a whiff of it every time you walk past, and the dense growth is easily pruned into beautiful standards and shapes if you want to try your hand at topiary. There are also trailing forms that will tumble beautifully out of a container or down a wall.

The secret to keeping rosemary happy and healthy is good drainage. A drought-tolerant plant, it will rot out and die if exposed to excess moisture, especially in the winter.

If you live on the edge of the hardiness zone, look for the cultivar 'Arp', which is cold tolerant, and place it in the warmest, best-drained spot in your garden. If it gets too cold for your rosemary to overwinter outside, it adapts beautifully to life in a container and can be overwintered indoors. The best condition for overwintering indoors is a sunny, cool spot where temperatures stay below 60°F (15°C).

Pruned into Christmas tree shapes, rosemary can sometimes be sold as a houseplant around the holidays, but it really doesn't survive well in modern, well-heated houses.

↑ Making the harvest. ↗ Rosemary makes a great partner for a grapevine and a clump of strawberries. ↑ The thick, woody, upright stems of rosemary.

PURPLE FLOWER RASPBERRY

Rubus odoratus

Layer: Shrub

Edibility: Fruit

Light requirements: Shade

Size: 4 to 5 feet (1.2 to 1.5 m) tall and wide

Habit and form: Suckering, deciduous shrub

Flowering and fruiting times: Flowers all summer, followed by berry-like fruit

Native range: Eastern North America

Lowest hardiness temperature: −40°F/−40°C

Ecological benefit: Host plant for many caterpillars, fruit valuable for birds

Pest and disease: None serious

Propagation: Cuttings or lift and divide suckers

Pollination requirements: Self fertile

Raspberries are delicious. With their rangy growth habit and vicious thorns, however, they're usually not a good addition to an ornamental landscape.

A few kinds of upright-growing raspberries are available that make a beautiful, and tasty, addition to your garden—and won't make the space a danger zone. The best of these is the purple-flowering raspberry. It grows as a thicket of narrow, thornless stems covered with large, attractive, maple leaf–shaped leaves. Starting in early summer and continuing through fall, the stems will be topped with large, pink, fragrant flowers that then develop into small, tasty red raspberries. This plant prefers shade, so adding it is a great way to bring both flowers and edible fruit into difficult garden spots. It will sucker and spread into a large colony over time, so either plant it in a space where it has room to roam or prune it back to contain it.

↑↑ Purple flowering raspberry are striking specimens. ↖ The plant produces sweet purple blooms. ↗ The delicious red fruit.

The Herbaceous Perennial Layer

The herbaceous perennial layer is where you can have the most fun. These plants are smaller than shrubs, so you can pack many of them in even the smallest garden. There are also many species and varieties to choose from, with lots of edible options. Proper perennial choice can round out the aesthetics of your garden and ensure that you have edible things to harvest throughout the year.

One of the best things about perennials is that, unlike trees and shrubs, they're easy to dig up and move, rearrange, or even give to a friend if you don't like them. A tree is a commitment that often lasts decades, but you can change all your perennials in the next season, so this is a layer where you can experiment, explore, and really play.

There are so many edible perennials in this photo, along with a mixture of plants from other layers, too.

The herbaceous layer includes a variety of plants with different shapes, sizes, heights, and sun/moisture needs. From full sun to shade (left to right): garlic chives, anise hyssop, buck's-horn plantain, red-veined sorrel, mint, ostrich fern, giant butterbur.

Remember that, even if a plant is a perennial, that doesn't mean it will live forever. Some herbaceous perennials (think peonies [*Paeonia* spp.], rhubarb, and asparagus) will live for decades, possibly outliving you. Others—like strawberries or cardoons—tend to live a few years and then fade away. So plan on replacing some of your perennials over the years, either by replanting the same thing or taking the opportunity to try something new.

Me, in front of my lupins (the seeds of which are edible if processed properly) and a few other edible perennials.

HOSTA

Hosta spp. and hybrids

Layer: Herbaceous perennial

Edibility: Leaves

Light requirements: Light to full shade

Size: Varies by cultivar, from a few inches (5 to 6 cm) to 2 to 3 feet (0.6 to 1 cm) tall and wide

Habit and form: Clumping perennial with bold leaves

Flowering and fruiting times: Flowers ranging from white to purple in mid to late summer

Native range: Asia

Lowest hardiness temperature: −40°F/−40°C

Ecological benefit: Fill in difficult shady spots where few plants thrive and flowers provide nectar for bees and hummingbirds

Pest and disease: Slugs and deer

Propagation: Lift and divide plants in early spring before they come into growth

Pollination requirements: NA

Hostas, native to Japan, Korean, and China, are beloved garden perennials you can find nearly everywhere. They are appreciated for their bold leaves, which will thrive in the difficult shady areas of your garden, but they can also be grown as a vegetable.

The species *Hosta montana* and *Hosta seiboldiana* are harvested most often as vegetables in Japan, but all species and hybrids of hostas are edible. The largest, most vigorous forms will give you the biggest harvests, but explore the many available hybrids to find the ones that taste best to you.

Once the leaves have fully unfurled, they're too tough to be good eating. The best time to harvest them is when the tightly rolled leaves are just poking out of the ground like spikes. For a truly gourmet version, cover the emerging shoots with an upturned pot to keep out all light and harvest them like white asparagus, as is done in Japan. The flavor is somewhere between lettuce and spinach, and they can be eaten raw (if they're nice and tender) or sautéed, boiled, roasted, or, as is traditional in Japan, breaded and fried as tempura.

Harvesting the hosta's leaves weakens them, so only take a few shoots from each plant. If you have a very large hosta that's taking over too much garden space, this can be a great way to keep it in check and give you delicious vegetables at the same time.

↖ These hosta shoots are ready for harvest. ← Breaded in tempura batter and fried is one of the best ways to enjoy hosta shoots.

SOLOMON'S SEAL

Polygonatum odoratum 'Variegatum'

Layer: Herbaceous perennial

Edibility: New shoots

Light requirements: Light to full shade

Size: 18 inches (45 cm) tall, spreading indefinitely

Habit and form: Spreading rhizomatous perennial with arching stems

Flowering and fruiting times: Small white flowers in spring

Native range: Asia

Lowest hardiness temperature: −40°F/−40°C

Ecological benefit: Fill difficult spots, pollinators attracted to the flowers

Pest and disease: None significant

Propagation: Lift and divide plants, ideally in early spring before they come into growth, but can be successfully divided most times of the year

Pollination requirements: NA

The genus *Polygonatum* is a large one, with many species native to different parts of the world, but the variegated *Polygonatum odoratum* 'Variegatum' is by far the most common in gardens, and for good reason. It is a remarkably durable plant for shaded conditions, thriving in dry or wet soils, laughing at cold and heat, and generally going about its business without problems. It spreads rhizomes to make an ever-increasing patch over time, but it doesn't spread rapidly enough to become a pest.

Aesthetically, the great feature of Solomon's seal is its bright white variegation on the margins of the leaves, which can bring touches of brightness and light to areas in deep shade. The small white flowers dangling below the stems are attractive but not particularly noticeable.

In the kitchen, use the fresh new stems as they emerge in the spring, while they're still young and tender. These can be cooked and enjoyed as you would asparagus. Harvest these young shoots in moderation, as overharvesting will harm the plant. You should also avoid harvesting more than a quarter to a half of the new shoots, unless the plant is beginning to spread too much for your garden space; if that's the case, overharvesting can be a great way to limit its spread and make room for your other garden plants.

↗ Solomon's seal makes a great partner for edible hostas in shady spots. → This photo shows the best stage of growth for harvesting Solomon's seal shoots in the early spring. →→ Mature variegated Solomon's seal plants in a shade garden.

GARLIC CHIVE
Allium tuberosum

Layer: Herbaceous perennial

Edibility: Leaves and flower buds

Light requirements: Sun to shade

Size: 12 to 18 inches (30 to 45 cm) tall, 12 inches (30 cm) wide

Habit and form: Bulb with grassy foliage

Flowering and fruiting times: Flowers in late summer

Native range: China

Lowest hardiness temperature: −40°F/−40°C

Ecological benefit: Late summer nectar source for bees and other beneficial insects

Pest and disease: None serious

Propagation: Lift and divide the bulbs in the fall

Pollination requirements: NA

↗ The pretty white blooms of garlic chives. ↗↗ In this image, garlic chives are living in harmony with sweet potato vines. ➔ A harvest of garlic chive flower buds.

There is so much to recommend garlic chives in the garden. They'll grow happily in a wide range of climates and soils, thriving and blooming in sun or shade, and produce big clouds of white flowers in late summer when often the garden is going through a quiet moment.

Better still, they're delicious. They are a popular vegetable in their native China. If you've never tried them, the name says it all. Like chives, you can eat the leaves and flowers—the flowers being traditionally harvested in bud before they open—with a flavor more like garlic than oniony chives. The leaves can be harvested at any time in the summer, so it's easy to pick a few and chop them to add a quick garlic flavor to any dish.

OSTRICH FERN
Matteuccia struthiopteris

Layer: Herbaceous perennial

Edibility: Newly emerging fronds

Light requirements: Shade

Size: 3 feet (1 m) tall, 2 feet (60 cm) wide

Habit and form: Tall upright fern

Flowering and fruiting times: NA

Native range: Much of the Northern Hemisphere

Lowest hardiness temperature: −40°F/−40°C

Ecological benefit: Only host plant for the Ostrich Fern Borer moth

Pest and disease: None serious

Propagation: Dig up and move offsets

Pollination requirements: NA

↗ Ostrich ferns spread quickly and form a large colony. ↘ Mature ostrich ferns. ↘↘ Early spring is when the fiddleheads are harvested for consumption.

Ferns are a staple of the shaded landscape. One of the oldest groups of plants to evolve, they've stuck around because they thrive and grow in a wide range of conditions. Ostrich ferns have big, dramatic, upright-growing fronds. They grow taller in wet conditions, and a little shorter in dry, but will thrive in a wide range of shaded garden conditions.

Be aware that, especially in the wetter conditions they love, ostrich ferns can spread aggressively, so they might not be the best choice for very small garden spaces. One way you can slow them down is by harvesting their fronds to eat.

The right stage for harvesting is early in the spring, when the new fronts are emerging but still curled tight in a beautiful roll called a fiddlehead. Cook them completely before eating—steaming is the recommended method—and enjoy. And note that nearly any fern can be eaten as a vegetable, though not all are equally tasty. Be sure to cook bracken fern (*Pteridium aquilinum*) completely before eating to remove a toxic compound it contains.

ANISE HYSSOP

Agastache foeniculum

Layer: Herbaceous perennial

Edibility: Leaves and flowers

Light requirements: Sun

Size: 3 feet (1 m) tall, 2 feet (60 cm) wide

Habit and form: Upright clumping perennial

Flowering and fruiting times: Flowers summer through fall

Native range: Central North America

Lowest hardiness temperature: −40°F/−40°C

Ecological benefit: Flowers attract pollinators, the many tiny seeds offer a valuable food source for birds during the colder months if the sturdy seedheads left intact

Pest and disease: Root rot in wet soils

Propagation: Divide in the spring or from seed

Pollination requirements: Self fertile

Anise hyssop is a beautiful, extremely drought-tolerant plant. It blooms with spikes of many tiny purple flowers starting in the summer and carrying on right into the end of fall, especially if deadheaded as the flowers fade. The individual plants can be a little short lived, but when happy it will self-sow to perpetuate itself.

Every part of the plant has a strong, delicious anise or licorice smell and flavor. Its leaves can be used to make tea or flavor desserts and the fragrant flowers are a beautiful and delicious addition to salads or can be used to garnish a beverage. Harvesting the flowers has the added benefit of encouraging rebloom, so don't be shy about enjoying them.

↗ At the back of this photo, you'll see purple edible Agastache blooming in my garden. The orange-flowering Agastache at the front of the photo is also edible, but it is typically not as flavorful. → A harvest that's ready to be dried and turned into tea. →→ Anise hyssop (*Agastache foeniculum*) has strong licorice-flavored foliage.

LEMON VERBENA
Aloysia citrodora

Layer: Herbaceous perennial

Edibility: Leaves

Light requirements: Sun to part shade

Size: 3 feet (1 m) tall, 3 feet (1 m) wide

Habit and form: Woody shrub that dies back to the ground each winter

Flowering and fruiting times: Flowers in summer

Native range: South America

Lowest hardiness temperature: 10°F/–12°C; grown as an annual in colder zones

Ecological benefit: Flowers attract pollinators

Pest and disease: None serious

Propagation: Cuttings

Pollination requirements: NA

Crush a lemon verbena leaf and you'll know exactly how it got its name. The lemon smell is strong and sweet. It's a pleasure to brush up against this plant in the garden, and adding a few stems will elevate the aroma of any flower arrangement. The leaves are edible as well, used as a flavoring or, commonly, to make a bright, lemony tea. You can use it to pump up the flavor of lemonade or add a bright lemony note to desserts or savory dishes.

Where the temperature doesn't drop much below freezing, lemon verbena will grow into a large, woody shrub. In colder climates, it will die back to the ground each winter and come back in the spring, acting like a herbaceous perennial. If your temperatures regularly drop below 15°F (9.5°C), either grow it as an annual or dig it up and bring it inside for the winter.

← The blooms of lemon verbena. ↑ Lemon verbena plants growing harmoniously with perennial sage in the garden of a friend.

CARDOON

Cynara cardunculus

Layer: Herbaceous perennial

Edibility: Base of leaves

Light requirements: Sun

Size: 3 feet (1 m) tall and wide

Habit and form: Vase-shaped perennial

Flowering and fruiting times: Flowers around midsummer

Native range: Mediterranean

Lowest hardiness temperature: 0°F/–18°C; grown as an annual in colder zones

Ecological benefit: Flowers attract a wide range of pollinators and the seedheads are a great food source for birds

Pest and disease: Root rot in wet soils

Propagation: Grow from seed and from division

Pollination requirements: Self fertile

Cardoons are the more durable, less fussy cousins of artichokes. As an ornamental, they have a lot to recommend them, with huge, dramatic silvery leaves topped with enormous purple flowers in midsummer. They're living sculptures for the garden, thriving in sun and well-drained soils; they laugh at drought.

The edible part of this plant is in the base of its leaves, sort of like celery. The entire plant must be cut to harvest, as the inner leaves are the most tender. Trim away the leaves and tough stems, then simmer the bases of the stems in water with lemon juice in it for thirty minutes to tenderize and remove any bitterness. The result will be a vegetable that tastes much like an artichoke. While they're also like artichokes in that they can be a lot of work to prepare, cardoons are deliciously worth all the labor.

↗ Cardoon plants growing in the garden. ↘ The edible leaf stalks of cardoon. ↘↘ Cardoon makes a bold and striking plant in the garden. This one is surrounded by a ground cover of thyme.

MINT

Mentha spp.

Layer: Herbaceous perennial

Edibility: Leaves

Light requirements: Sun to part shade

Size: 18 inches (45 cm), spreading indefinitely

Habit and form: Spreading perennial

Flowering and fruiting times: Flowers in June

Native range: Europe and Asia

Lowest hardiness temperature: −40°F/−40°C

Ecological benefit: Bees love the flowers

Pest and disease: Spider mites can be a problem in hot, dry conditions

Propagation: Divide or root cuttings

Pollination requirements: NA

Mint is delicious. With its bright flavor, it adds a wonderful zing to herb salads, lemonade, and, of course, a mint julep.

As a garden plant, mint is easy to grow, generally free of pests and disease, and tolerant of a wide range of conditions. The clusters of small purple flowers are pretty and attractive to pollinators. In fact, the biggest problem with mint is that it grows *too* well. Happy mint plants send out long runners that quickly colonize huge sections of the garden and are almost impossible to eliminate. If you have a big area, you can let it run, or grow it in a large container.

↗ Mint can be quite aggressive in the garden. Give it lots of room or plant it in a container. ↘ Mint shoots ready for harvest. ↘↘ A harvest of fresh mint ready for making tea.

ASPARAGUS
Asparagus officinalis

Layer: Herbaceous perennial

Edibility: New shoots

Light requirements: Sun

Size: 3 to 5 feet (1 to 1.5 m) tall and wide

Habit and form: Clumping perennial with lacy, fernlike foliage

Flowering and fruiting times: Flowers in early spring, females have small red berries in late summer

Native range: Europe

Lowest hardiness temperature: −40°F/−40°C

Ecological benefit: Bees feed on pollen from the flowers

Pest and disease: Generally low maintenance, but can have fungal diseases, especially when stressed

Propagation: Lift and divide plants in early spring before coming into growth or grow from seed

Pollination requirements: Separate male and female plants

Asparagus needs no introduction as a vegetable: It's common in grocery stores and on restaurant menus. In the spring, the spears of new growth shoot up from the ground, are cut while still young and tender, lightly cooked, and enjoyed.

But asparagus is also a vegetable that ticks all the boxes for an edible layered garden. A perennial that lives a long time, it sends out root systems that search deep in the soil for moisture and nutrients. After you harvest the first flush of shoots, the next should be left alone so that the plant can grow and photosynthesize. That's when the familiar asparagus transforms into a strikingly beautiful, airy cloud of delicate, fernlike foliage.

In this form, asparagus serves to visually fill spaces in the garden, as well as providing a contrast to large, bold-textured plants. Asparagus comes in separate male and female plants. Since the berries aren't eaten, males are generally preferred as a vegetable. But if you do have a mix of male and female plants, the females will produce small, red berries in summer that just add to the beautiful effect in the garden.

Asparagus can be grown from seed or purchased as young plants. You'll have to wait a few years after planting for them to get established before you start harvesting.

← Purple asparagus is a fun way to add some variety to your garden. ↖ New asparagus shoots ready for harvest. ↗ Asparagus is planted as dormant crowns.

RED-VEINED SORREL

Rumex sanguineus

Layer: Herbaceous perennial

Edibility: Young leaves

Light requirements: Sun to part shade

Size: 12 inches (30 cm) tall and wide

Habit and form: Clumping perennial with large showy leaves

Flowering and fruiting times: Flowers in summer

Native range: Europe, southwest Asia, and northern Africa

Lowest hardiness temperature: −20°F/−29°C; can be grown as an annual in colder climates

Ecological benefit: NA

Pest and disease: None serious

Propagation: Grow from seed

Pollination requirements: Wind pollinated

↗ The red veins of sorrel make quite a statement in the garden.
↘ Red-veined sorrel is growing in a window box with edible pansies.
↘↘ Get ready for a tart but flavorful punch.

The name tells you what to expect here: broad green leaves decorated with attractive dark red veins. The leaves are beautiful any time of the year, making a great addition to containers or in-ground plantings.

Red-veined sorrel can be easily grown from seed, and nurseries will sometimes have it for sale. But remember that, if you find this in the ornamental, rather than edible, section of the nursery, it may have been treated with chemicals that aren't food safe. Ask before purchasing and, if necessary, wait for new growth to flush out before harvesting.

The leaves of sorrel have a tart lemony taste, and the young, fresh leaves are the most tender and enjoyable. As the plant matures, the leaves get larger and tougher. When grown as a vegetable, the whole plant is often harvested while young, before the leaves get tough, but you can also cut just a few and let the plant mature to keep enjoying the ornamental foliage. Trimming the plants back at the end of summer will encourage a crop of new tender leaves in the fall.

BUCK'S-HORN PLANTAIN

Plantago coronopus

Layer: Herbaceous perennial

Edibility: Leaves

Light requirements: Sun to light shade

Size: 4 inches (10 cm) tall and wide

Habit and form: Low growing rosette

Flowering and fruiting times: Flowers in midsummer

Native range: Europe, Asia, and North Africa

Lowest hardiness temperature: −20°F/−29°C

Ecological benefit: Birds feed on the seeds; host plant for the buckeye butterfly caterpillar

Pest and disease: None serious

Propagation: Grow from seed

Pollination requirements: Primarily wind pollinated

This easy-to-grow little plant might become your new favorite green. It grows as a rosette of attractive, narrow, slightly lobed leaves, topped in summer with little spikes of less showy green flowers. The plants are durable, cold hardy, and drought tolerant, though they'll produce the most tender and delicious leaves when grown in richer conditions.

You can harvest the leaves anytime, though they are the sweetest in the cooler months of the year. The leaves have a distinctive crisp, crunchy texture and a delicious nutty flavor. Toss them raw into any salad or, when they may have gotten bitter in the summer, cook them as you would any other green.

↑ Buck's-Horn plantain is best harvested in the spring for fresh eating. ←← If you're harvesting in the summer, cooking the greens is the best option to reduce their bitterness. ← The plants are cold hardy and drought tolerant.

DAYLILY

Hemerocallis spp. and hybrids

Layer: Herbaceous perennial

Edibility: Flowers, new shoots, and tubers

Light requirements: Sun to light shade

Size: Depends on variety, generally 2 to 3 feet (0.6–1 m) tall and wide

Habit and form: Clump of grassy foliage

Flowering and fruiting times: Flowers in midsummer, often into the fall

Native range: Asia

Lowest hardiness temperature: −40°F/−40°C

Ecological benefit: NA

Pest and disease: Deer, occasional rust

Propagation: Divide in the spring

Pollination requirements: NA

Daylilies are essential components of any garden because they're so easy to grow, can adapt to many different conditions, and produce big, gorgeous flowers. Modern daylily breeding has taken this diverse genus and transformed it into a bewildering array of forms and colors. You can get daylilies that are taller than you, or tiny ones that will barely reach your knee. Some have huge, heavy blooms that are ruffled and patterned with different colors, and others make clouds of small, delicate, lemon-scented blooms. There is truly a daylily for every gardener. And they're all edible—you can eat nearly every part of the plant.

The fresh, tender new growth in the spring can be harvested and cooked much like spinach. The leaves will grow tough as they mature, so harvest only the youngest shoots and peel away the outer leaves to get the tender, inner layers.

The roots of daylilies also produce small tubers, which are edible. These are best harvested in the late fall to early spring when they have the most stored carbohydrates. During the summer, the tubers will be mushy and not worth harvesting. You can dig a plant, cut off some of the tubers, then replant it to ensure future harvests.

When harvesting both shoots and tubers, take no more than half from each plant to make sure it will continue to thrive and produce. And only harvest from the most vigorous varieties. Usually the old-fashioned hybrids and species, like the ubiquitous orange *Hemerocallis fulva*, are the best options for these types of harvests.

With fancier, slower-growing, or more expensive daylilies, stick to harvesting the flowers, which can be picked without harming the plant.

Fully open blooms have a crisp texture and mild flavor when eaten raw, much like lettuce, and are a beautiful addition to any salad. The darkest blooms tend to have a stronger, slightly bitter flavor, while lighter colors, like yellows, are often the tastiest.

You can also harvest the whole flower buds the day before they would open. These can be breaded and fried, or dried and then added to soup.

↑ Flower buds of daylilies can be harvested, breaded, and fried. ↖ Open flower petals can be eaten as-is or whole blooms can be breaded and fried. ↗ Daylily tubers are best harvested in late fall to early spring.

VIRGINIA BLUEBELL

Mertensia virginica

Layer: Herbaceous perennial

Edibility: Stem, leaves, and flowers

Light requirements: Shade

Size: 2 feet (60 cm) tall and wide

Habit and form: Ephemeral spring perennial

Flowering and fruiting times: Spring

Native range: Eastern North America

Lowest hardiness temperature: −40°F/−40°C

Ecological benefit: Early pollen source for bees

Pest and disease: None serious

Propagation: Divide after the plants go dormant

Pollination requirements: NA

↗ The stems, leaves, and flowers of Virginia bluebells are edible. ↗↗ Harvest the shoots in early spring, when they are 4 to 8 inches (19 to 29 cm) tall. ↘ Never pick more than a few stems from each plant in any given year.

Virginia bluebells are a spring woodland ephemeral: They emerge early in the spring, grow and flower fast, then go dormant as the tree canopy above them comes into leaf. The wide, attractive leaves and clusters of dangling blue flowers are familiar garden plants, but they are also an excellent vegetable.

The best time to harvest is when the shoots are four to eight inches (10 to 20 cm) tall. Cut the entire stems, including leaves and flower buds, but only take a couple stems from each plant. A quick steam or sauté is all that's needed to cook and transform them into a delicious green. Once you taste it, you wonder why you haven't been eating bluebells for years and will look forward to their short, delicious, harvest period every spring.

GIANT BUTTERBUR, GREEN FUKI

Petasites japonicus

Layer: Herbaceous perennial

Edibility: Leaf, stems

Light requirements: Shade

Size: 2 to 3 feet (0.6 to 1 m) tall, spreading widely

Habit and form: Wide leaves from spreading rhizomes

Flowering and fruiting times: Unique clusters of white flowers on stubby stocks in early spring before the leaves emerge

Native range: Asia

Lowest hardiness temperature: −20°F/−29°C

Ecological benefit: Early pollen source for bees

Pest and disease: None serious

Propagation: Division

Pollination requirements: NA

Butterbur has large leaves that make beautiful textural contrast in a shady garden, especially when paired with finer-textured plants.

The dramatic foliage looks positively tropical, if not Jurassic. This plant loves to grow in wet conditions, so if you have a soggy, shady corner of the garden, it's the perfect choice. Be aware, however, that it can spread quite rapidly, so give it room to run. Before planting, make sure giant butterbur is not listed on your region's invasive species list.

In Japan, the butterbur is called fuki, and serves as a popular, wild-harvested vegetable. The petioles—stems of the leaves—are harvested, not the leaves themselves.

Properly cooked, these stems are delicious, but they do need to be cooked well or they'll taste bitter. Precook them by boiling for five minutes in salted water, then soak them in a richly flavored broth of your choice. The result will be crisp, savory, and delicious.

↑ The edible stems of fuki need to be prepared properly. ↗ The large leaves of this plant are striking in the garden, but be aware that the plant spreads quickly.

The Climber Layer

Many gardens are missing a climber layer. Walk through your neighborhood and you'll probably see every other layer mentioned in this book, represented somewhere in nearly every yard—but rarely will you find climbers.

This is unfortunate, because climbers can be a perfect way to fill the empty spaces in a tight urban garden. Every building on your property has empty walls that could be covered with beautiful and delicious climbers. You can also create structures specifically for climbers, which is a great way to divide areas, screen out ugly views, or provide shade over a sitting area.

Scarlet runner beans supported by a trellis and backed by blueberry bushes are a favorite in the climber layer.

Climbers

In addition to covering built structures, climbers can grow up trees and shrubs. This can add another harvesting opportunity for an existing tree, but use caution and match the vigor of the climber to the size of the tree. A delicate, nonedible clematis (*Clematis* spp.) can drape over a medium-sized shrub without doing it much harm, but an edible, hardy kiwi (*Actinidia arguta*) will smother and potentially kill all but the largest canopy trees with their massive vines.

Different climbers use different methods to climb, which determines how they will behave in the garden. Adherers have roots that stick to surfaces. English ivy (*Hedera helix*) is a classic example, also one that's terribly aggressive and invasive (and inedible), but Boston ivy (*Parthenocissus tricuspidata*) and Virginia creeper (*Parthenocissus quinquefolia*) are others that are not as aggressive (but also not edible).

These types of climbers will zip up any surface, be it the side of a building or a tree trunk. Choose their locations with care, as they can be very hard to remove once they have attached to the side of your garage. Older brick buildings can also be damaged by clinging roots, though new construction uses stronger mortar that shouldn't be affected. In general, climbers do less damage to the trees they climb because they cling to the trunk, rarely shading out the healthy branches of the tree. They can, however, add weight to a tree and contribute to its failure (this is particularly true of English ivy).

Twiners

Twiners are a subclass of climber that wrap around things to support themselves as they climb. They may wrap their whole stem around something, as kiwi, hops (*Humulus lupulus*), and runner beans (*Phaseolus coccineus*) do, or put out little tendrils to wrap around supports, like peas or grape vines.

This will have a few key impacts in the garden. Twiners can't adhere to a smooth wall or fence but will need some sort of trellis or net to wrap themselves around. Plants with smaller tendrils, like peas, require more slender materials like mesh, string, cable, or thin bamboo stakes, whereas more robust plants like kiwi can wind themselves around larger materials like wooden stakes.

These grapes are trained along a fence and surrounded by a ground-covering layer of thyme.

This can be a huge advantage, because a twiner put up the side of your house won't grow beyond the edges of the trellis you erect, and they can be easily cut down and removed without damaging the wall behind them. Be sure to introduce twiners to trees and shrubs with care, though, as they tend to climb up over the small outer branches and twigs and can shade out and kill the plants they grow over. You can grow them on trees and shrubs, but be ready to get out your pruners if they become too aggressive.

Leaners

The final group in the climber layer are the leaners. Climbing roses and some raspberries and blackberries are the best example of these types of climbers. In some sense they aren't climbers at all: They don't cling to surfaces or branches, but rather have evolved to thrive in dense thickets of vegetation, sending out long stems that can lean and drape through and over the plants around them. They often have recurved prickles that help them catch onto other branches.

In a garden setting, leaners can be allowed to grow up and through small shrubs, or they can be grown up walls, trellises, and arbors, but you, the gardener, will have to manually tie the long, flexible stems to their supports. This can be nice, as you have total control over where they grow, but remember that it only takes one growing season of neglect for your beautifully trained climbing rose to turn into a wild thorny mess.

Plan for Your Climbers

Whatever the type of climber you choose, the key to success is planning. Because climbers rely on the plants and structures around them for support—instead of growing strong stems themselves—they can grow extremely fast.

Do your research on how big any climber you might plant will get, plan ahead for any pruning you'll need to do to keep it in check, and ensure that you build structures strong enough to support the mature vine. Some of the best edible vines, like kiwis, get massive and heavy with time. Plan on building a robust structure for these large vines. Believe me, you don't want to deal with cleaning up a heavy vine that's pulling down a flimsy arbor.

Though they aren't profiled in this book, American groundnuts (*Apios americana*) are a member of the climber layer that produces an underground edible tuber. It's also known as the Indian potato.

KIWI

Actinidia arguta and *A. deliciosa*

Layer: Climber

Edibility: Berries

Light requirements: Sun

Size: To 20 feet (6 m), depending on trellis size

Habit and form: Woody deciduous vine

Flowering and fruiting times: Flowers in early spring, fruit ripens in late summer

Native range: Asia

Lowest hardiness temperature: −40°F/−40°C (*A. arguta*); 10°F/−18°C (*A. deliciosa*)

Ecological benefit: Nectar source for bees

Pest and disease: None serious

Propagation: Cuttings of named forms

Pollination requirements: Separate male and female plants required for pollination

↗ Hardy kiwi fruits are not covered in brown fuzz like the the more familiar species. ↘ The fruits are small but sweet. ➜➜ Attractive kiwi flowers on the vine.

The kiwis you see in the supermarket develop on fast-growing vines that are perfect for covering trellises or training up a wall. *Actinidia deliciosa* is the familiar fuzzy species you've seen at the store. They're only hardy where temperatures don't regularly drop below 15°F (−9°C). In colder climates, you can try *Actinidia arguta*, which is extremely cold tolerant, growing happily where winter temps drop as low as −40°F (−40°C). *A. arguta*'s fruits have no fuzz on the outside and are a bit sweeter than the traditional kiwi fruit, but they tend to be smaller. If you don't have room for the two plants normally required for pollination, look for the variety 'Issai', which is self fertile and therefore doesn't need a companion to give you a good harvest.

Whichever kiwi you grow, remember that the plant will grow into large, heavy, woody vines. Build a structure for it to grow on accordingly, as a flimsy trellis will eventually collapse under the plant's weight. It will not climb up a flat wall without some sort of trellis, but it will be fine twining around built supports or even trees.

MAYPOP
Passiflora incarnata

Layer: Climber

Edibility: Fruit

Light requirements: Sun to part shade

Size: To 10 feet (3 m), depending on trellis size

Habit and form: Deciduous vine

Flowering and fruiting times: Flowers throughout the summer followed by fruit

Native range: Southeastern North America

Lowest hardiness temperature: −20°F/−29°C

Ecological benefit: Host plant for the Gulf fritillary butterfly

Pest and disease: None serious

Propagation: Cuttings, division

Pollination requirements: Self fertile

Maypop is the common name for one of the really cold-hardy species of passion vine. If you live in a frost-free climate, you can grow beautiful, delicious passion vines (*Passiflora edulis*), the main source for passionfruit), but in colder climates, *Passiflora incarnata* should be your choice.

This vigorous, often suckering vine, has classic, almost over-the-top passionflower blooms, followed by small green fruits. Some fruits will drop on the ground when they're ripe and ready to be eaten, while others will cling to the plant. These you can leave until their green rind turns yellow.

The skin is tough and not pleasant to eat, but inside ripe fruits will have a yellow flesh around black seeds (if you see white flesh and seeds, the fruit isn't ripe). You can eat the pulp fresh, spitting out the seeds, or press it through sieve to remove the seeds and make delicious jam.

As beautiful as maypop passion vines are, they can spread aggressively, both with their above-ground shoots and in long runners they send underground, so choose a spot where it will be easy to contain it, or grow it in a wild area where you can let it run.

↗↗ Maypops are aggressive growers. Be sure to have a supportive structure for them to climb. ↑ Passionvine flowers. ↑ This passionfruit is ripening, but not yet ready for harvest.

RUNNER BEAN

Phaseolus coccineus

Layer: Climber

Edibility: Pods, seeds, flowers

Light requirements: Sun

Size: To 10 feet (3 m), depending on trellis size

Habit and form: Annual vine

Flowering and fruiting times: Flowers all summer, followed by edible pods

Native range: South America

Lowest hardiness temperature: 10°F/–12°C; grow as an annual in colder climates

Ecological benefit: Flowers attract hummingbirds

Pest and disease: None serious

Propagation: Seed

Pollination requirements: Self fertile

Runner beans are the close relatives of the familiar green bean, but they offer an ornamental twist, with large, showy red (or white in some cultivars) flowers that are pollinated by hummingbirds. The flowers are incredibly beautiful, and then are followed by tasty bean pods that you can harvest and cook exactly like green beans. These pods will get large, but pick them when still young and tender for the best eating experience.

You can also let the pods fully mature and harvest the mature beans from them. They have beautiful, dark seeds flecked with dramatic pink, and can be cooked and enjoyed like any other bean in your favorite stew or chili.

Much like other beans, they're best grown by planting the large seeds directly in the garden once all chance of frost has passed and the soil has warmed up above 60°F/15.5°C. Runner beans come by their common name thanks to their rapid and rampant growth habit. They climb by wrapping their stems around structures. Be sure to sure to provide a large trellis for them to run up.

↑ The mature beans are flat and flavorful. You can eat them freshly cooked like regular green beans or allow the pods to dry on the vine and harvest the dried beans for eating. ➔ Runner bean plants are quick climbers.

HOPS
Humulus lupulus

Layer: Climber

Edibility: Bracts around female fruits

Light requirements: Sun

Size: To 30 feet (9 m), depending on trellis size

Habit and form: Deciduous vine

Flowering and fruiting times: Flowers in summer

Native range: Asia

Lowest hardiness temperature: −30°F/−34°C

Ecological benefit: Host plant for question mark, comma, and red admiral butterflies, among others

Pest and disease: None serious

Propagation: Cuttings

Pollination requirements: Only female plants typically grown, pollination not required

Hops are primarily grown to flavor beer. The long, vigorous vines die back to the ground each winter, then shoot up into growth in the spring. The flowers are small catkins, and on female plants these develop into leafy structures that look like pinecones, with bracts around the developing seeds. Those bract structures are harvested to lend their distinctive taste and aroma to beer. Without pollination, female plants won't have fertile seeds, but they still produce the bracts and flavor, so it's typical to only grow females.

Since hops grow fast, they're a great way to screen out an unpleasant view or make a space more private. Simply put up a trellis and watch it disappear. For added ornamental impact, look for golden hops (*Humulus lupulus* 'Aureus'), which has bright golden-yellow foliage in spring that slowly fades to green as the season progresses.

↑ Hops harvest. ↗ Hops vines are aggressive growers and spreaders. Give them ample room and a supportive structure.

GRAPES
Vitis vinifera

Layer: Climber

Edibility: Fruit, leaves

Light requirements: Sun

Size: To 30 feet (9 m), depending on trellis size

Habit and form: Deciduous woody vine

Flowering and fruiting times: Flowers in early spring, fruit ripens late summer

Native range: Europe

Lowest hardiness temperature: −30°F/−34°C; variety dependent

Ecological benefit: Fruits consumed by wildlife; host plant for Abbott's sphinx moth caterpillars, among others

Pest and disease: Numerous fungal diseases, especially in humid climates

Propagation: Cuttings or grafting of named forms

Pollination requirements: Self fertile

Grapes are vigorous vines that can get enormous, but they can also be pruned back to control their size. One of the best ways to enjoy them in the home garden is to train them over the top of a pergola or sitting area, giving you shade when they are leafed out in the summer, with their hanging clusters of fruit. In the winter they drop their leaves to allow you to enjoy a little sun.

There are any number of grape varieties to enjoy, but for eating purposes, remember that some varieties are specifically bred to be made into wine, while others, called table grapes, are selected to have tender skins and are ideal for eating fresh. Which you choose will depend on how you wish to use them. In addition to the typical European grapes, there are also other species and hybrids with other species that offer a little more disease resistance and interesting flavors in the fruit.

A few bonuses when adding a grapevine to your garden include the grape flowers. Though tiny, green, and visually unremarkable, they have a delicious fragrance, which makes sitting under an arbor in spring a magical experience. Another, less widely known benefit: Many grape vines have wonderful fall color, blushing orange and red, especially after a frost.

↑ There are many different varieties of grapes. Look for a combination of disease resistance and flavor. ↗ Proper pruning and training is essential for healthy grape growth.

The Annual Layer

The main focus of the layered edible garden is on perennial crops, for all the benefits of reduced work and soil disturbance we've discussed throughout this book. But annual crops definitely have their place as well.

You may want to include some traditional vegetables, like tomatoes or peppers, simply because you love to eat them, and annual crops also serve as temporary space fillers and weed blockers to put between shrubs and herbaceous plants that haven't yet reached their mature size. You can choose big sprawling squashes and zucchinis to temporarily cover large areas, or low-growing, nitrogen-fixing bush beans, which will increase the fertility of the soil as they break down at the end of the season.

Annual crops can also be a great option to fill empty spaces in *time*. Most of your herbaceous layer will be dormant in the winter, leaving a bare and inedible garden. Depending on how cold your winters are, you may be able to plant cold-tolerant crops like lettuce, kale, or mustard greens in the fall and harvest them in the spring to clear the space for the perennial plantings to fill in.

In general, annual crops need richer soil and more moisture than their perennial counterparts, because they have to put on a lot of growth very fast. So enrich the soil for your annual plantings with some extra compost and plan to put them where they can be reached easily by your hose.

Because annual crops are so common and easy to find on the market, I won't include separate profiles of them here. The collage of images on the following pages offers some great examples of less common annual crops to include in your layered edible garden.

The annual layer includes many common vegetable garden plants, including bush beans, zucchini, tomatoes, and more.

This garden consists mainly of the annual and root crop layers, but it also makes use of the climber layer, too.

A Sample of Unusual but Delicious Annual Edibles Worth Exploring

↑↑ Cabbages and kales ↖ New Zealand spinach ↗ Corn salad or mâche

↑↑ Strawberry spinach ↗↗ Orach ↖ Mizuna ↗ Ground cherries

The Ground Cover Layer

The lowest layer of the garden is the easiest one to overlook, but it may be the most important. Small, low-growing, carpeting plants play a key role in protecting the soil from erosion and summer heat while also eliminating places for weed seeds to germinate. Generous mulch can take its place if necessary, but a living mulch of ground covers can be maintained more easily once it's established and will help provide another layer of habitat and food for you and your garden's wildlife.

The ground cover layer is essential for reducing maintenance in your layered garden. It blocks weeds, prevents soil moisture loss, and so much more.

The most common choice for this layer is a lawn. Regardless of how we define a lawn, even including the traditional high-maintenance turf grass, you may have a place for such a feature in your layered garden. This is especially true if you need a durable ground cover for children or pets to play on. For other areas, though, there are many options that don't require mowing, irrigation, or chemical inputs in the form of fertilizer or pesticides. Rethinking exactly what you need from each space in your landscape can open up creative opportunities for new, interesting ground covers.

Foot traffic is a key consideration for ground cover choices. Some will tolerate frequent walking and playing—though even turf grass will struggle under excessively high traffic, as occurs with paths. Other ground covers, like thyme, will tolerate light, occasional foot traffic, and still others, like strawberries or ground cover raspberries, tolerate no foot traffic at all. One solution for a high-traffic path is to add stepping stones to reduce the weight placed on the ground cover, then plant more delicate, interesting ground covers around the stones.

When planting ground covers to fill in a large area, it can be hard to know how many plants to use. Most ground covers will spread indefinitely, in theory at least, so just a few plants can be used to fill in a large space given enough time. The more time it takes to fill in, though, the longer you will have to look at bare soil, which you'll have to weed.

Look at how far a given ground cover will spread each year, and aim to space your plantings so that they'll cover a space in one or two years. You can go for a wider spacing if your budget is tight and you have adequate time to weed, but keep the space weeded while your plants establish themselves. Cover the bare ground with mulches like leaves or straw, which will break down into the soil by the time the ground cover plants reach the bare spots.

In my garden, I grow a combination of ground covers, including self-heal, strawberries, and thyme, to name just a few.

This shady nook is home to bunchberries (*Cornus canadensis*), an edible ground cover that thrives in low light.

Because they are low growing, ground covers are generally easily shaded and killed by tall weeds. Extracting established, perennial weeds from mature ground cover is a huge challenge, even more so when the weeds are thorny. Success with establishing the ground cover layer depends on investing time in the initial weeding, with a particular focus on any longer-lived weeds that spread via rhizomes—they're the hardest to tackle and can continue to spread underneath your ground cover.

Successful ground covers also take into account the soil and conditions in which you are growing them. Put the wrong ground cover in the wrong spot and it will grow too thin to effectively cover the ground and prevent weed germination. For example, a creeping thyme will make a beautiful, edible carpet in a dry, sunny spot, but it will fail if planted in an area that's moist and shaded.

If you do your research and choose the right ground cover for your space—and are extremely vigilant pulling weeds while the ground cover gets established—you'll reap the reward of minimal work later and for years to come.

Another aspect of ground covers to bear in mind: While they're small and easy to watch over, don't ignore the fact that many ground cover plants are wonderful producers of edible fruits and offer a terrific habitat for native insects.

They're the secret tool to radically increasing the diversity of your garden, making less work for yourself, and getting more food out of a small space.

WILD STRAWBERRY

Fragaria virginiana, F. vesca, or *F. chiloensis*

Layer: Ground cover

Edibility: Fruit

Light requirements: Full sun to light shade (full sun for best fruiting)

Size: 6 inches (15 cm) tall, spreading 6 inches (15 cm) per year

Habit and form: Clumping plants spreading and multiplying by runners

Flowering and fruiting times: Blooms in spring, with fruit ripening in early summer

Native range: *Fragaria virginiana* in North America, *F. vesca* in Europe, *F. chiloensis* on the Pacific Coast of the Americas

Lowest hardiness temperature: −30°F/−34°C

Ecological benefit: One of the top host plants for butterfly and moth caterpillars (see below), fruits eaten by birds and small mammals

Pest and disease: Few problems if you choose a species native to your region; hybrid cultivars are prone to numerous fungal diseases

Propagation: Dig up and separate plants produced at the end of the runners

Pollination requirements: Hybrid cultivars are self fertile, but some wild species have separate male and female plants; if you want to collect seeds, always plant multiples

↗ Strawberries with creeping thyme make an easy-to-grow ground cover. ↘ The fruits of alpine strawberries. →→ Ducks can help keep strawberries free from slugs.

While hybrid cultivated strawberries can be garden divas, their wild ancestors are vigorous, easy-to-grow ground covers. Weave them in between the perennial edibles in your borders or mix them with grasses and sedges as a lawn alternative for low-traffic areas. Different species of wild strawberries are native to nearly every region of the world, so choose your local native form and they'll be a great addition to your garden.

Wild strawberry fruits are tiny compared to what you're used to finding at the grocery store, but they far surpass the big hybrids when it comes to flavor and fragrance.

The only thing that will love your strawberries more than you do are local butterflies and moths. In my region, over seventy different butterfly and moth species use strawberries as food for their caterpillars, and there are similar numbers anywhere strawberries are native. Some species that rely on strawberries include the common swift and the emperor moth in Great Britain, the brown-tail moth in Europe and parts of Asia, and the grizzled skipper and two-banded checkered skipper in the Americas, among many others. Small native bees and other pollinators also find this plant useful for providing nectar, and many species of birds enjoy the fruits.

Most strawberries spread by runners—horizontal stems that root and grow new plants at their tips—making them easy to propagate. The runners allow wild strawberry plants to find their own way into gaps in your garden and fill them in. If you have a very small garden space or want to keep your strawberries contained, look for the runnerless Alpine strawberries, a form of *F. vesca*, which produce fragrant, tiny berries all summer long and stay put as low clumps rather than running.

GROUND COVER RASPBERRY

Arctic raspberry, *Rubus arcticus* × *stellarcticus*, and creeping raspberry, *Rubus rolfei*

Layer: Ground cover

Edibility: Fruit

Light requirements: Full sun to light shade (full sun for best fruiting)

Size: 6 inches (15 cm) tall, spreading 6 inches (15 cm) per year

Habit and form: Long stems spread along the ground, rooting as they go

Flowering and fruiting times: Blooms in spring, with fruit ripening in early summer

Native range: Arctic raspberry, cold regions of the Northern Hemisphere; creeping raspberry, Asia

Lowest hardiness temperature: −50°F/−18°C (Arctic raspberry); −10°F/−18°C (creeping raspberry)

Ecological benefit: Host plant for many caterpillars, fruits eaten by birds and small mammals

Pest and disease: Rabbits and deer will browse the leaves

Propagation: Dig up and separate rooted stems

Pollination requirements: Grow two or more varieties to ensure good fruit set

The genus *Rubus* includes blackberries, raspberries, and other fruits, most of which grow into tall, thorny thickets that can be a little difficult to fit into a home landscape. But there are some species that grow flat against the ground, making the perfect ornamental ground cover, with the added bonus of providing delicious berries to eat. Arctic raspberries are some of the best for this, with new varieties created by hybridizing two different species native across the Northern Hemisphere. The dense foliage is topped with pink flowers in the spring, transitioning to red berries in the summer, and then, as an added bonus, the foliage turns brilliant shades of crimson and burgundy in the fall.

As you might guess from the name, Arctic raspberries grow best in climates with cool summer temperatures and are some of the most cold-tolerant fruiting plants you can grow. In hotter climes, you might consider creeping raspberry, which has the same low-spreading growth habit, but with foliage that stays evergreen through the winter, and sweet berries shaded a beautiful apricot-yellow color.

↑ A ready-to-plant Arctic raspberry. ↗↗ Arctic raspberry in bloom.
↗ The small but sweet fruits of Arctic raspberries.

WILD GINGER

Asarum spp.

Layer: Ground cover	

Edibility: Rhizomes

Light requirements: Light shade to full shade

Size: 5 inches (15 cm) tall, spreading 3 to 4 inches (7–10 cm) per year

Habit and form: Broad leaves from spreading rhizomes

Flowering and fruiting times: Flowers in early spring

Native range: Different species native over much of the Northern Hemisphere

Lowest hardiness temperature: –30°F/–34°C

Ecological benefit: Unusual flowers are food source for pollinating flies and beetles

Pest and disease: None

Propagation: Dig and divide the spreading rhizomes

Pollination requirements: NA

There are numerous species of wild gingers. If you're in Europe, it would be *Asarum europeanum*; in North America, *A. caudatum*, *A. canadense*, or *A. speciosum*; and in Asia, a host of beautiful species including *A. maximum* and *A. splendens*. Wherever you are, and whichever wild ginger you grow, these plants offer durable, easy-to-grow ground covers that thrive in deep, dry shade and resist damage by deer, rabbits, and other pests.

Most species are evergreen, with glossy leaves often marked with silver. Asarum flowers are beautiful as well, but may require you getting down on your knees to enjoy them. Most species are pollinated by beetles, so the small, usually brown flowers open down at ground level where they're easy to overlook.

The name wild ginger refers to the gingery smell and taste of every part of the plant. The underground stems (rhizomes) have the most intense flavor. To harvest, gently dig up your plants in the spring. You'll find a network of rhizomes connecting individual clumps of leaves and roots. Cut off the sections of rhizome between the clumps and replant. The plants will carry on growing unharmed, and you can use the harvested rhizomes the way you would ginger to make a tea or to spice a pumpkin pie.

← While wild ginger is not the same as the ginger you get with your sushi, it does have a ginger-like flavor. ↖ Wild ginger forms a thick mat of leaves. ↗ The small underground rhizomes of this harvested ginger can be used to make tea or flavor a pumpkin pie.

WINTERGREEN

Gaultheria procumbens

Layer: Ground cover

Edibility: Leaves and berries

Light requirements: Light shade to full shade

Size: 3 to 6 inches (7 to 10 cm) tall, spreading 3 to 4 inches (7 to 10 cm) per year

Habit and form: Short stems with evergreen leaves spreading to form a carpet

Flowering and fruiting times: Flowers in summer, berries from fall through winter

Native range: Eastern North America

Lowest hardiness temperature: –30°F/–34°C

Ecological benefit: Berries are food source for many animals in the winter

Pest and disease: None serious, occasional mildew or leaf spot

Propagation: Dig and divide the spreading rhizomes

Pollination requirements: Self fertile

↗ Wintergreen plants are very cold tolerant and produce wintergreen-flavored foliage. ↗↗ Wintergreen in flower. ↘ Wintergreen.

The name wintergreen refers to the thick, glossy green leaves of this plant that look absolutely impeccable every month of the year, even in the middle of winter. The short stems are topped by small, bell-shaped white flowers in midsummer, and these are followed by bright red berries that persist through the winter. In the garden, one of the best features of this plant is its ability to grow in deep shade where few other plants will thrive. It grows best in acidic soil with ample organic matter.

You already know the taste and smell of wintergreen: We usually experience it flavoring a breath mint or chewing gum. These days, most wintergreen flavoring is artificial, but you should grow some to experience the real thing. The red berries look beautiful and have a faint wintergreen flavor, but they're not sweet or appealing. The real prize here is the foliage, which has long been popular as a flavoring and in tea. Unlike most teas, wintergreen releases the most flavor when allowed to steep a long time, so fill a jar with leaves, cover it with water, and let it sit for three to five days. The flavored liquid can then be drunk as tea or used to make wintergreen sorbet.

One warning: Wintergreen contains a natural chemical called methyl salicylate that converts to salicylic acid when consumed; this is the active ingredient in aspirin. This means that wintergreen tea can have some of the same pain-relieving qualities as aspirin, but, as with aspirin, it can be toxic in large quantities. Homemade wintergreen tea has low concentrations of methyl salicylate, but wintergreen oil, a highly concentrated extract, should be treated as medicine rather than flavoring.

THYME

Thymus spp.

Layer: Ground cover

Edibility: Leaves

Light requirements: Full sun

Size: 2 to 3 inches (5 to 8 cm) tall, spreading 3 to 4 inches (7 to 10 cm) per year

Habit and form: Low spreading carpet of fragrant leaves

Flowering and fruiting times: Flowers in early summer

Native range: Mediterranean Europe

Lowest hardiness temperature: −20°F/−29°C

Ecological benefit: Great nectar source for bees

Pest and disease: Root rot in wet soils

Propagation: Dig and divide or root stems tips as cuttings

Pollination requirements: Self fertile

↗ A mixed planting that includes thyme. → The colorful flowers of thyme are appealing to many pollinators—these are my favorite, *Thymus longicaulis* (Mediterranean creeping thyme). →→ There are many species and varieties of thyme you can grow in your garden, though some are better for eating than others. These thymes are grown alongside the larger-leaved silvery lamb's ear (*Stachys byzantina* 'Silver Carpet').

You probably have dried thyme in your spice drawer, or have cooked with fresh thyme leaves. As delicious and flavorful as thyme is in the kitchen, it's equally wonderful in the garden. There are many species and selections of thyme, and in general they thrive in full sun and well-drained soil. They're also great choices for hot, dry spots where other plants might suffer from drought.

Thymus vulgaris is the standard edible species. Though delicious, it isn't the most ornamental of the genus, as it does not flower heavily and has an open, slightly ragged growth habit. Creeping thyme (*Thymus serpyllum*) makes a tighter, more attractive carpet of leaves, blooming heavily in the spring. It's also edible, with a slightly more minty flavor than traditional thyme.

Lemon thyme (*Thymus citriodorus*) is another terrific plant that's edible and highly ornamental, making for a tight, beautiful ground cover. It's delicious in cooking, affording a bright lemon note on top of its traditional thyme flavor. For even more ornamental impact, look for the variegated version of lemon thyme, which tastes the same, but has a narrow cream edge to each leaf. My favorite species of creeping thyme is Mediterranean creeping thyme (*Thymus longicaulis*), as it's a reliable and handsome evergreen ground cover in most areas and, in my experience, is less susceptible to root rot in wet winter conditions.

SELF-HEAL

Prunella vulgaris

Layer: Ground cover

Edibility: Leaves

Light requirements: Full to part sun

Size: 12 inches (30 cm) tall (when in bloom), slowly spreading clump

Habit and form: Upright clumping perennial

Flowering and fruiting times: Flowers from summer to fall

Native range: Europe, Asia, Africa, and North America

Lowest hardiness temperature: −30°F/−34°C

Ecological benefit: Great nectar source for bees

Pest and disease: None serious

Propagation: Dig and divide or root stem tips as cuttings; also easy from seed and will self-seed

Pollination requirements: NA

Self-heal, also called heal-all, is a plant native to much of the Northern Hemisphere. Nearly everywhere in its native range it has been eaten as a vegetable and used as a healing herb. The small leaves have a mild, tasty flavor, similar to lettuce, and can be eaten raw or cooked. They're most tender—and most suitable for eating raw—in the spring as the leaves emerge. In the summer they're tougher and more palatable if cooked before eating.

Prunella is also a great ornamental plant because it starts flowering around June, with short spikes of purple (or pink, occasionally white) flowers that keep on coming right through fall. The flower display is beautiful, much loved by bees, and also edible. Sprinkle some flowers over a simple salad to transform it.

↗ Self-heal in bloom. ↘ The flowers are enjoyed by many pollinators. You can enjoy them, too, as they're edible. ↘↘ A self-heal harvest.

CREEPING OREGON GRAPE

Berberis [formerly *Mahonia*] *repens* or *B. nervosa*

Layer: Ground cover

Edibility: Berries

Light requirements: Sun to shade

Size: 1 to 2 feet (30 to 60 cm) tall, spreading 6 to 8 inches (15 to 20 cm) per year

Habit and form: Low-spreading evergreen shrub

Flowering and fruiting times: Flowers in early spring, fruit in late summer

Native range: North America

Lowest hardiness temperature: −20°F/−29°C

Ecological benefit: Hummingbirds and bees nectar at the flowers

Pest and disease: Leaf spot, and leaves will scorch in winter in exposed sites

Propagation: Dig and divide

Pollination requirements: Self fertile

↗ This holly-leaved shrub produces edible flowers and berries. ↗↗ Oregon grape berries ready for harvest. ↘ The yellow flowers of Oregon grape.

Creeping or low Oregon grapes have glossy, evergreen foliage, very showy clusters of yellow flowers early in the year, and are followed by equally attractive clusters of silvery blue berries. These durable, low shrubs spread slowly to form a ground cover. The leathery evergreen leaves are spiny and similar to holly, so they're best placed away from frequently traveled paths to avoid scratchy interactions. Cold, windy weather can lead to scorching discoloration on the leaves, so they're best sited in a spot sheltered from winter wind.

Both the flowers and berries are edible, the flowers a cheery addition to a salad. The berries are intensely sour in flavor, so best to use these as an ingredient in cooking rather than eating them raw.

If you have space for a larger plant, you can also consider *Mahonia aquifolium* (tall Oregon grape), which has all the same virtues in the form of a taller shrub, reaching three to ten feet (1 to 3 m).

NASTURTIUM

Tropaeolum majus

Layer: Ground cover

Edibility: Leaves and flowers

Light requirements: Sun to light shade

Size: 6 inches (15 cm) tall to 12 inches (30 cm) wide

Habit and form: Trailing annual

Flowering and fruiting times: Flowers all summer

Native range: South and Central America

Lowest hardiness temperature: 20°F/–7°C; grow as an annual

Ecological benefit: Nectar source for bumblebees

Pest and disease: None

Propagation: Grow from seed

Pollination requirements: NA

↗ The colorful edible blooms of nasturtiums. ↘ Nasturtiums ramble around other plants in the garden, forming a thick and protective ground cover. ↘↘ Edible nasturtium bloom.

Nasturtiums are easy-to-grow annuals. The large seeds quickly germinate and grow into vigorous, spreading plants with attractive green leaves and an abundance of showy flowers in shades of yellow to red. As annuals, they're a great choice to fill in empty areas around newly planted shrubs or perennials, covering the ground and outcompeting weeds while the permanent plantings get established.

Most varieties are clumping forms, but if you have a big space to fill, look for the climbing, which have longer stems that can cover an area of a few feet (1 m). They can also be tied up to cover a fence or trellis, but they won't climb on their own.

All parts of the plant are edible, offering a sharp, peppery flavor. The flowers and youngest leaves have the mildest flavor and are great additions to salads, while the more mature leaves are stronger. The immature seeds can be pickled to use as an alternative to capers.

EASTERN PRICKLY PEAR

Opuntia humifusa

Layer: Ground cover

Edibility: Pads and fruit

Light requirements: Sun

Size: 1 foot (30 cm) tall, spreading to 2 feet (60 cm) wide

Habit and form: Flat-paddle cactus

Flowering and fruiting times: Flowers in early summer, fruit ripening in late summer

Native range: North America

Lowest hardiness temperature: −30°F/−34°C

Ecological benefit: Nectar source for bees, thrives in dry sites where little else can grow

Pest and disease: None

Propagation: Individual pads placed in the garden will root and grow into new plants

Pollination requirements: Two or more different varieties required for fruit set

↗ Winter-hardy in even cold climates, prickly pear makes a dense and edible ground cover. ↘ Both the pads and the fruits of prickly pear are edible. ↘↘ These prickly pear fruits are ready to harvest.

Prickly pears are the most adaptable and easy-to-grow cactus in nearly any climate. While many cactus are limited to warm, dry climates, the eastern prickly pear is native to a wide swath of North America and will thrive even in places with rainy summers and cold winters.

The pads of prickly pears are called nopales when eaten as a vegetable and have a flavor reminiscent of green beans. Harvest newly grown pads in the spring, as older ones will be tough and fibrous.

The most important step is to clean them. The pads can have large spines and, in addition, small hair-like spines called glochids. Retaining either will ruin your meal. To clean the pad, cut off its outer edge, then lay it flat and rub down the surface with a sharp knife or vegetable peeler. When you have gotten all the clusters of brown glochids off, rinse with water and wipe down with a towel to make sure all the hairs have been removed gone. After that, the pad can be eaten raw or cooked.

The fruit is edible as well. It's tart and sweet, but also covered with hairs. To get at the flesh inside, cut off the ends of the fruit, slice it in half, and scoop out the center of the fruit, discarding the spine-covered skin.

The Root Crop or Rhizosphere Layer

When you look out at your garden, the branches, stems, and leaves that you see are only half the story—just as much is going on underground in the root systems of your plants. The primary function of roots is to absorb water and nutrients and provide support for the top half of the plant, but many plants also store energy in their roots and tubers. They produce thick, expanded roots and underground stems filled with starch and other nutrients that the plant can then draw on to fuel growth in other seasons.

Plants that store energy underground are called "geophytes." Their underground storage organs can be modified leaves (bulbs), stems (rhizomes, tubers, or corms), or roots. It so happens that many of these underground stores of nutrients are delicious. You know carrots, potatoes, and onions (a root, a tuber, and a bulb, respectively), but there are hosts of other plants that have edible tubers and roots.

Traditional root crops include carrots, radish, and parsnips.

Working an edible rhizosphere into your landscape takes some planning, as these plants need to be dug up to harvest. Choose locations that you can access easily, such as the perimeter of beds or the back of a border. Don't place rhizomes close to trees with dense root systems that will be difficult to dig through. The top half of your rhizosphere crop, like the tall flowering stems of a sunchoke, will generally act as part of the herbaceous perennial layer, though quick-growing root crops like radishes are great additions to the annual layer.

LEFT AND ABOVE: Mashua plants (see page 189) are one of the most interesting members of the root crop layer. Their edible tubers make for a beautiful and delicious harvest.

SUNCHOKE
Helianthus tuberosus

Layer: Underground

Edibility: Tubers

Light requirements: Sun

Size: 6 to 8 feet (1.8 to 2.4 m) tall, 3 feet (1 m) wide

Habit and form: Herbaceous perennial

Flowering and fruiting times: Flowers in midsummer

Native range: North America

Lowest hardiness temperature: −40°F/−40°C

Ecological benefit: Pollinators, food for caterpillars

Pest and disease: None serious

Propagation: Division or seed

Pollination requirements: Requires two or more varieties for seed

Sunchokes are a tall, perennial sunflower that bloom with masses of sunny yellow flowers in midsummer. They will grow happily in a wide range of climates and conditions, tolerating drought and lean soils, though you'll get the best yield of edible tubers if they're grown in fertile soils with a regular water supply.

After the stems have died back in the fall, you can harvest the edible tubers. Lift the plant with a shovel or garden fork and you'll find a big mass of knobby tubers. You can harvest and eat most of these, replanting a few chunks of tubers to grow for next year. The tubers are delicious when cooked like potatoes, offering a satisfying, nutty, somewhat artichoke-like flavor (which is the source of their name).

Sunchokes contain high concentrations of a natural starch called inulin, which is delicious but does not get processed in our digestive tract, which means eating sunchokes doesn't cause insulin spikes the way a starchy potato might. Beneficial bacteria in our gut *do* feed on inulin, so it is considered a prebiotic, helping support a healthy gut microbiome. That all sounds good. However, if you aren't used to eating inulin, a big meal of it will cause extremely unpleasant amounts of gas. It's best to start with small amounts, slowly increasing how much you eat to allow your gut bacteria to adjust. You can also cook them in lemon juice or pickle them, as acid compounds will break down the inulin into more digestible forms of starch.

One further warning: When planted in the ground, sunchokes will reappear year after year if their stem tubers are left in the soil; eventually they will spread and take over. If you have a large garden, this might be just fine. But if you need to control them in a smaller space, consider growing your sunchokes in large containers or grow bags—something like 10 gallons (38 L) or larger is ideal to give the tubers room to form and to provide stability for the tall plants.

↑ My sunchoke harvest. ↗ Sunchoke plants grow tall and produce pretty yellow flowers. ↗ The plants are quick spreading, so give them plenty of room and make regular harvests.

OCA

Oxalis tuberosa

Layer: Underground

Edibility: Tubers

Light requirements: Sun

Size: 2 feet (60 cm) tall and wide

Habit and form: Herbaceous perennial

Flowering and fruiting times: Orange-red flowers in summer

Native range: Mountains of South America

Lowest hardiness temperature: 20°F/−7°C or dig up and store the tubers indoors for the winter

Ecological benefit: Great source of nectar for hummingbirds and pollinators, especially late in the season.

Pest and disease: None

Propagation: Tubers

Pollination requirements: NA

Appearing as an attractive clump of bright green, cloverlike foliage, oca has a long history of cultivation as a food crop in South and Central America. The short days of fall trigger the plant to start producing tubers, which will be ready to harvest in late fall or early winter. The tubers come in a number of bright colors, ranging from yellow to orange and red, depending on the cultivar. They can be eaten raw or cooked and enjoyed much like a potato, though they have a slightly tart flavor that is all their own. After harvesting, be sure to save a few tubers to replant next year in the spring.

Oca is beautiful, easy to grow, and delicious in climates with cool summers, but it struggles in hot summer climates, especially in humid areas where temperatures don't drop lower than the 70°F (20°C) at night. If your climate has early frosts in the fall, it may also be difficult to get a good crop. Oca is best grown in climates like the Pacific Northwest in America, Western Europe, New Zealand, and its native western regions in South America. If your climatic conditions don't match these places, you might try another root crop from that region, mashua (*Tropaeolum tuberosum*), which has large, edible roots and grows as a large vine. In addition to the edible roots, mashua has attractive, edible, nasturtium-like leaves and starts blooming late in the year with extremely showy red and yellow flowers. Those late blooms are a great source of nectar for pollinators. In my garden, the Anna's hummingbirds love them!

↑ Harvested oca tubers. ↗ Oca foliage.

QUAMASH

Camassia spp.

Layer: Underground

Edibility: Tubers

Light requirements: Sun to light shade

Size: 18 inches (45 cm) tall, 12 inches (30 cm) wide

Habit and form: Bulbous perennial

Flowering and fruiting times: Spring to early summer

Native range: Western North America

Lowest hardiness temperature: −30°F/−34°C

Ecological benefit: Pollinators attracted to the showy flowers

Pest and disease: None

Propagation: Divide bulbs

Pollination requirements: NA

You'll find camassias for sale in nearly every fall bulb catalog. They're great additions to the garden with tall spikes of beautiful blue flowers. Even if you don't eat them, you'll be happy to add them to your garden. The bulbs are edible, and you can harvest them in the fall, replanting some to grow next year.

The most common recommendation for preparing camassias is to peel them and cook them long and slow in an oven or a pressure cooker until they turn golden and the indigestible starches break down and turn sweet. Then they can be mashed or fried like potatoes.

Camassias were an important food source for indigenous North Americans, who had developed techniques for identifying the bulbs, harvesting them, and cooking them, all in a sustainable way to allow the wild populations to continue to thrive.

If you garden in the native range of camassias, be careful to avoid the death camas, *Toxiocoscordion venenosum*, which produces bulbs that look similar but are poisonous. It's easy to tell apart in flower, as the death camas has spikes of yellowish-white blooms, unlike the blue of camassias.

In your garden, there will be no confusing the two as you will know what you've planted, but *never* harvest camassias from natural populations without being absolutely certain what you are harvesting.

↑ A field of quamash. ←← Quamash flowers. ← Camassia tubers.

TARO
Colocasia esculenta

Layer: Underground

Edibility: Corms

Light requirements: Sun or shade

Size: 3 feet (1 m) tall and wide

Habit and form: Herbaceous perennial

Flowering and fruiting times: Usually doesn't flower in cultivation

Native range: Eastern Asia

Lowest hardiness temperature: 0°F/−18°C; grow as an annual in colder climates

Ecological benefit: NA

Pest and disease: None

Propagation: Division

Pollination requirements: NA

→ Taro plants. ↘ Freshly harvested taro plants and corms. ↘↘ The interior of a taro corm.

Taro, usually called elephant ear when sold as an ornamental, has huge, tropical-looking leaves. In climates with hot summers, it's vigorous and easy to grow. It will thrive in full sun or shade, and in average to wet soil conditions. You'll get the biggest yields of edible corms in full sun and consistently irrigated soils. There are ornamental forms with dark purple or yellow-green leaves that are just as edible as the plain green form, but be aware that many other plants are sold under the common name of "elephant ear," so double-check that what you're buying is *Colocasia esculenta*. Corms sold at international grocery stores will grow just fine if you put them in the garden.

Taro is a staple vegetable in South Asia, Oceania, and parts of Africa. Cook the starchy corms much as you would a potato or sweet potato.

5

MAINTENANCE

A layered edible garden is less work than a traditional vegetable garden, but there are still regular maintenance tasks you'll have to plan for to ensure your garden stays healthy and productive.

VIEW MAINTENANCE AS part of the pleasure of having a garden. You get to work in the space you've created and spend time making it more pleasant and function better. This can give you a great connection to your garden and the natural rhythms of the world around you, and it can be a way to spend quality time with friends and family.

The key is to plan the amount of work your garden will require so that it matches the time you have to give to it. The quickest way to ruin a garden—and your gardening experience in general—is to make it so much work that you resent it or simply can't keep up with it. Think through the maintenance that will be required as you plan your garden, and don't be afraid to admit if it's getting to be too much and you need to scale back the garden for a while or hire professionals to help with the larger tasks.

Your garden should enrich your life, not be another chore to deal with.

HARVESTING

Harvesting hardly qualifies as a task for some people because it's the fun part, the payoff for all the hard work and planning. Nevertheless, it's something you need to plan and set aside time for, particularly with plants that produce edible fruit. If you don't harvest edible leaves, shoots, or roots, they'll just carry on growing and can be harvested later—either later in the season, or in the next year if the food is only edible for a short period. But fruit, if not harvested, can drop and become a mess, attracting pests like rats or aggressive hornets. While smaller fruits and berries are cleaned up by birds if you don't harvest them, so they're not usually a problem—though even small fruit can be messy and unpleasant if it drops on a patio or the sidewalk—letting larger fruits like apples drop and rot on the ground can lead to a number of headaches.

Providing climbing structures or various tall plants for vine crops to ramble up is one small part of taking care of your layered edible garden.

Colorful and flavorful harvests await you!

With this in mind, be sure to plan time in your schedule to harvest your crops. It's best to place heavy fruit producers in spots of the garden where they won't cause a problem if you miss some of the crop and it ends up dropping to the ground.

PRUNING

Maintaining trees and shrubs will often involve pruning. The seemingly simple act of cutting back your plants with shears or loppers can have a huge impact on how they grow, and feeling comfortable with the basics of pruning can help you keep your garden safer, healthier, and more productive.

Not all trees and shrubs need to be pruned. There are a few reasons you'll reach for the shears: increasing fruit or flower production, maintaining the vigor and health of plant, removing hazards for safety, or controlling and directing new growth.

Pruning Woody Plants by the Seasons

Spring

PROS:
- Pruning in the spring triggers vigorous flushes of new growth.
- Sub-shrubs (woody plants like rosemary, thyme) and suckering shrubs respond well to spring pruning.
- Early spring is a great time for rejuvenating more mature plants, since many will respond by sending out fresh growth.
- If you want to encourage plants to bush out and provide more new shoots and leaves, pruning in spring is the ideal time.

CONS:
- Blue light from the sun at this time promotes vigorous vegetative growth, so avoid pruning in spring if the goal is to maintain a small and tidy shape.
- Also avoid pruning in early spring if you get late cold snaps: If your plant pushes out new growth shortly after pruning, it can be severely damaged by the cold.

Summer

PROS:
- Summer is a great time to reduce the size of woody plants.
- Pruning in summer typically will not trigger a flush of new growth.

CONS:
- Stronger light, especially in the ultraviolet band, can burn foliage or parts that were previously shaded. To protect your plants, avoid pruning on intensely sunny days.
- Late summer pruning can trigger some new growth, which is not good as a plant prepares to enter chilly weather in fall. This can damage tissues that haven't had time to harden off.
- Many fruit trees produce their flower buds in summer, so late pruning can remove those buds, reducing flowering and the fruit set for next year.

Fall

PROS:
- Great for maintaining proper shape of shrubs and hedges before entering the winter/snowy season.
- This is a good time to shape formal hedges to look sharp through the winter, also to remove branches that could be damaged by heavy winter snows.
- It's also a good time to prune conifers and hedges.
- Since the growth response isn't as vigorous in the fall, you won't get as much accelerated growth.

CONS:
- Not ideal for certain climates that remain warm and sunny in the fall, as growth response may still be triggered, pushing out new growth that may not be thrive as you head into winter.

Winter

PROS:
- There will be less material (that is, leaves) to handle when the branches are bare.
- You can walk around trees more freely since herbaceous plantings are dormant.
- You can pursue greater gardening activity in the winter, collecting branches to use in next year's garden to make trellises and other support structures.
- Winter is also a great time for pruning conifers and many deciduous trees and shrubs (including fruiting ones).
- Most fruit trees, like apples, are best pruned in late winter.

CONS:
- Avoid pruning during cold snaps, as you don't want to expose new wounds to cold weather.
- Prune fruiting trees with care to avoid removing too many dormant flower buds.

Pruning Fruit Trees

Increasing fruit or flower production is the main reason you would prune most fruit trees.

Pruning fruit trees is a special skill unto itself, and the full details are beyond the scope of this book. You can find detailed instructions for the specific fruit trees you want to grow, but there are several basic principles you should keep in mind.

The main task is to remove or limit branches that will produce mostly leaves and direct more of the tree's energy into producing flower buds, which will develop into fruit. What this looks like will depend on the specific plant you're pruning.

On an apple tree, you will remove the stems called waterspouts—tall, vigorous shoots that grow straight up and produce only leaves. On raspberries, this will be removing all stems older than two years, as older stems no longer produce fruit. As you add fruiting trees and shrubs to your garden, you'll need to do a little research on how to properly prune each one to maximize fruit production.

Many fruit trees will try to produce more fruit than a tree can support. Apples are famous for this. If all the developing fruit is left on a tree, it will be so overwhelmed that the next year it will produce virtually no fruit, leading to a cycle called biennial bearing.

Thinning the fruit produced by the tree each spring while it is still small ensures a good harvest every year. All that fruit can also be heavy, so thinning also prevents damage to the tree's branches. Wild-type trees, which bear smaller fruit, tend not to have this problem in the same way experienced by highly bred forms with large fruit.

One of the biggest mistakes you can make while pruning a fruit tree—or other fruiting plant—is cutting off all the flower buds. For most fruit trees, flower buds are produced in the fall and sit dormant through the winter before bursting into growth in the spring. This means that fall-to-spring pruning should be done with care to avoid removing the buds. You can often tell the difference between flower and leaf buds by their appearance: Flower buds will be larger, fatter, and rounder when compared to leaf buds. Another way to avoid damaging flower buds is to prune immediately after a plant has flowered, before the new flower buds appear.

Pruning is crucial for maintaining the health and vigor of your plants. Even if you want to let your plants grow in their natural form, you need to watch out for problems. Winter is often the best time to find and cut out these problem spots—it's a time when plants have fewer leaves, revealing any issue that might otherwise be hidden by growth.

The Three Ds of Pruning

The first thing to look for when pruning are the three Ds: dead, diseased, or damaged parts. Cutting these branches out with a precise, clean cut will help prevent disease and other problems from spreading. Be sure to check in with your plants after extreme weather or heavy snows to remove damaged branches promptly, reducing the chance of further damage.

Next, look for branches that are growing back into the center of the plant or crossing over other branches, especially if they're rubbing against branches. Pruning these out will allow for better airflow through the plant, which helps prevent disease and lets more light in. You should always remove a branch that is rubbing against another one, as the rubbed area will create a wound that can be an entry point for disease.

Even more critical is pruning to remove safety hazards. Any woody plants near walkways, driveways, buildings, or patios need to be evaluated to ensure they're not growing where they'll become a problem. The sooner you can intervene here, the better. As a branch increases in size, it will be more work to prune and redirect. The obvious dangers here are large branches that could do damage as they fall, but also consider limbs hanging low enough for people to hit their heads. Additionally, anything sharp should be pruned well away from areas where people walk.

Different Pruning Options

Controlling and directing a plant's growth encompasses a huge range of different pruning options. How much of this pruning you do depends on your goals.

If you hate pruning and find it a chore, you can avoid nearly all of this by carefully choosing plants and siting them so their natural growth habits will work for you. If you find pruning an interesting exercise, or love fitting the most plants possible into a small space, these methods can give you a whole world to explore.

Bear in mind that the more extreme the pruning methods you take on here, the more work they will be to maintain and the quicker they will go downhill if you stop. Espalier pruning can transform a fruit tree into a nearly two-dimensional form that will fit in the narrowest of garden spaces, and sculpting topiary can transform a shrub into a dramatic piece of living sculpture. Trees pruned in this way will turn into a rangy mess in a single growing season without vigilant upkeep. So take on these kinds of pruning projects slowly and with an honest look at how much time you'll enjoy spending on them.

Proper pruning for fruit trees is essential for good production and healthy trees.

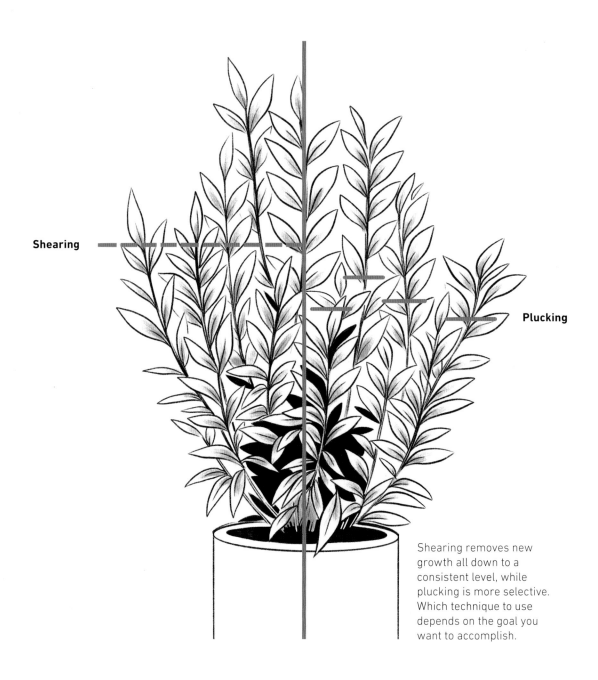

Shearing

Plucking

Shearing removes new
growth all down to a
consistent level, while
plucking is more selective.
Which technique to use
depends on the goal you
want to accomplish.

SHEARING AND PLUCKING

Shearing is one of the most common methods for
pruning a plant to maintain its shape. It entails simply
coming along and pruning off all the growth beyond a
certain predetermined shape, often as cubes or spheres.

Shearing gives a tight, formal look to a plant, and
can be quickly and efficiently done with large hedge
shears or power hedge trimmer. One rule you must
remember when shearing a hedge is to keep the base
of the hedge wider than the top. If the top gets wider,
it will shade out the bottom, leaving you a sparse,
open bottom. Never shear a hedge with a straight side;
instead, let the sides angle out slightly so they keep
getting sun. This is important on the north side of a
hedge (in the Northern Hemisphere), which will be
shaded more anyway.

Shearing does have some downsides, however. Because the sheared effect creates a tight, formal state, a plant will quickly look messy as new growth emerges, requiring more frequent shearing through the growing season to keep shrubs looking tidy. Cutting off all the new growth multiple times a season also means most sheared hedges will produce few, if any, flowers, fruit, or berries. So most shrubs that you grow for their edible fruits simply are not good candidates for shearing. On the other hand, if the edible part of a plant is the new growth, as with tea, shearing and harvesting can be the same action. Changing out a boxwood hedge for a tea hedge will give you the same formal, evergreen look, but with a regular harvest of homegrown tea.

If you want to limit the size of a shrub but keep a more natural growth habit and allow for the normal production of flowers and fruit, pluck pruning works beautifully. It is more labor intensive, requiring all cuts to be made with hand pruners rather than power tools, but this technique grows out more naturally, keeping the plant looking good even if you only do it once a year, or even once every couple of years for slower-growing plants.

Say you have a shrub that you want to keep at 3 feet (1 m) tall. With shearing, you draw a line at that height and cut off everything above that line and nothing below it. For pluck pruning, you select the branches that have grown the farthest above that line, then prune them back several inches below your imaginary line. Branches that are shorter than your line, or just a little beyond it, you leave in place. Those untrimmed branches look natural, and will be able to produce flowers and fruit as usual. In subsequent years, those untrimmed branches will grow longer, so you'll cut them back, while those you previously pruned will be able to grow out to maintain the natural look of the shrub, flower, and fruit. Follow the same rule as with shearing about keeping the base wider than the top to avoid shading the bottom branches out.

Arborizing

If you've inherited a shrub that has been pruned without keeping the base wider than the top, and the lower branches are thin and shaded out, you can transform them by a pruning practice called arborizing.

For this, you cut off all but a few of the lower branches, trimming them right back to one—or a few—trunks, and allow the top of the shrub to grow naturally. Essentially you transform a shrub into a subcanopy tree, showing off a beautiful trunk and branching structure and removing the tired-looking lower branches.

Arborizing also creates a space for shade gardening under the tree.

CROWN RAISING AND THINNING

Dealing with shade that's too deep from a large tree can be another reason to reach for the pruning shears. Deep shade can be a huge limiting factor in what you can grow in your garden. Though there are edible plants that will grow in shade, you'll have many more options if you can gain more light. You can, over time, alter dense shade into a brighter one by careful pruning.

This is by crown raising, the process of cutting off a few of the tree's lower limbs. Crown raising will lighten shade, but it doesn't create the most aesthetically pleasing effect and can damage the tree if done too aggressively.

A better option is often crown thinning, selective removal of branches in the crown to improve density while maintaining balance in the canopy. An arborist can also remove dead, diseased, and damaged wood for safety and let more light through.

Done properly, these methods will allow more light to penetrate the canopy and give you more options to grow something underneath. If you're just removing a few small limbs, this is a job you can do yourself. For larger jobs, hiring a licensed arborist will ensure it is done safely while maintaining the health of the tree.

The seedheads of many plants offer a food resource for songbirds in the winter, as well as offering an overwintering site for many different types of beneficial insects.

CUTTING BACK

Most herbaceous edible perennials will die back in the fall, leaving dead leaves and stems that stand over the winter before pushing out new growth in the spring. Cutting off that old, dead growth will give your garden a much cleaner, tidier look, and can help promote airflow and reduce disease problems.

Gardeners often cut everything back in the fall, but recently we've seen many benefits from embracing the different forms/stages of plants and waiting until the spring to cut back. The brown stems and especially seedheads that remain on a dormant tree can be attractive through the winter—certainly more so than bare soil—and many beneficial insects and pollinators hibernate in those dead stems and leaves through the winter.

Many birds depend on seedheads as well. Big hits with the birds in my garden include Echinacea (you can make tea from its roots), cardoon (with edible leaf petioles), and my favorite, anise hyssop. This last is great for tea and has sturdy stems and seedheads that persist beautifully through the winter months filled with hundreds of seeds for birds.

So the best option is usually waiting until spring, just before plants push their new growth and ideally when temperatures are consistently above 50°F (10°C) to allow overwintering insects to wake up.

The Three-Cut Method

Tree branches need to be cut properly to ensure that the cut heals fully and doesn't lead to damage for the tree.

First, identify the branch collar. This is a slightly raised or swollen section of wood right at the base of the branch where it joins the main trunk. All cuts should be made beyond this point. Cutting into the branch collar can harm the tree.

Cut 1: The bottom cut

A couple of inches (5 cm) or more, depending on size of branch, beyond the branch collar, make a cut from the bottom of the branch, about one third of the way through the branch. *This is very important:* If you skip this cut and start from above, the weight of the branch can tear the wood as it falls.

Cut 2: The top cut

Start this cut from the top, a little further away from the branch collar. Cut down to completely remove the limb.

Cut 3: Finishing

You'll now be left with a short stub of the branch. Cut this off from above, just beyond the branch collar. Paint or sealant is not required after removing the branch since, if the cut has been made properly, the branch collar can grow out and over the cut surface to seal the wound.

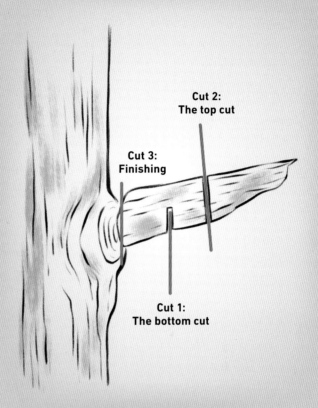

The three-cut method for properly removing tree branches with a pruning handsaw.

Pinching certain plants, like bee balm (*Monarda*) and culinary mints, means you'll have a better branching structure and more flowers per plant.

Blackberries and raspberries will need a trellis or some other type of support structure.

STAKING AND PINCHING

Some herbaceous plants will grow too tall and collapse in an ugly mess later in the summer.

There can be many reasons for this. Sometimes our breeding for large flowers or fruit creates something that's too heavy for a plant to support. Peonies are a classic example of this: The wild types, with just five petals, hold themselves up just fine, but the selected double forms have flowers packed with so many petals that they're too heavy—especially after a rain—for the stems to hold up.

Plants also start flopping over due to cultural conditions. Rich, highly fertile soil, and regular waterings encourage plants to grow tall and lush—sometimes much taller and lusher than they would in the wild, and that extra growth causes them to topple over. Finally, plants grown in a little too much shade will stretch tall to reach for the light and then collapse under their own weight.

There are a few solutions for flopping plants. You can try to change cultural conditions: Cut back waterings, don't add extra compost or fertilizer (especially nitrogen, which triggers lush growth), and move plants to brighter sun. You can also position plants so the things around them will provide support. Open shrubs are great sturdy options, and sometimes simply planting densely will allow all the plants to lean on each other.

Cutting back (pinching) summer- and fall-blooming perennials, including edibles like Monarda and anise hyssop, is a great way to prevent flopping. In early summer, as they are coming into growth, trim them back by one third. The trimmed stems will branch, maturing into a lower, more compact, less floppy version of the normal plant.

The final option is staking. You can purchase or make wire frames for the plants to grow through, or tie them to wooden, bamboo, or other stakes, or stick twiggy branches into the ground and allow the perennials to grow up through them.

Whatever technique you use, it's important that you do your staking before the plant starts falling over. Done early, the staking will be invisible and the plant habit will look normal. Once a plant has collapsed it is nearly impossible to wrestle it into an upright and attractive position. If a plant does collapse on you before you stake it, make a note on your calendar for next year to stake it sooner, or replace it with something that stands tall on its own.

DIVIDING

Clumps of herbaceous perennials, like chives, will expand each year, getting wider and thicker. Every three to five years you may want to consider lifting and dividing them in the spring before they flush out new growth. One reason to do this is to propagate them and make more plants to spread around the garden or share with friends, but there are also maintenance reasons to consider.

Some perennials, as they expand, will die out in the center, forming a doughnut of live growth with a dead patch in the middle. To rejuvenate these, dig up the clump, slice it into four or more pieces, cut out the dead center, and replant one of the live, happily growing clumps. The rest of the plant can be spread around the garden, given to friends, or just composted if you don't have a place for it.

In a densely planted garden, a large perennial may outgrow the space you have for it and end up competing with neighboring plants for space, sunlight, nutrients, and water. Digging it, dividing off a smaller chunk, and replanting will get it back into balance and keep it from encroaching on neighboring plants.

Some herbaceous edible perennials don't live long in nature: When left in one spot, they'll choke themselves out and start to decline. Regular division will help keep these plants thriving and producing year after year.

MULCHING

Mulch is simply a layer of dead organic matter (or even nonorganic material like rocks) covering the soil surface. This can be fallen leaves from trees, compost, the dead stems and leaves of last year's growth, pine needles, straw, wood or bark chips, shredded leaves—almost anything you can think of.

Keeping all of the bare ground in your garden covered with some kind of mulch has numerous benefits. It protects the soil from erosion, moderates cold and warm temperature swings, holds moisture in the soil to keep it from drying out, and significantly limits the ability of weed seeds to germinate.

In the layered garden approach, I advocate covering every bit of ground with some kind of plant.

Once your garden has matured, you'll have very little space to actually mulch. In a newly established garden, though, you'll have a lot more bare ground while plants are maturing, so keep that area mulched. You will probably need to renew the mulch annually, but this depends on your climate and mulching materials. Straw and shredded leaves break down faster than wood chip mulches, and everything decomposes faster in warm, humid climates than in areas that stay cooler or drier.

Check your mulches regularly and try to keep at least an inch or so of mulch over the soil surface, topping it up when the mulch breaks down.

Never pile mulches up over the crowns of perennials or up onto the trunks of trees and shrubs. This will hold in too much moisture and can cause rot and other diseases. Limit the mulch to the spaces around plants and keep a one- to two-inch (2.5- to 5-cm) gap between the base of plants and your mulch.

Though you can use nearly anything to mulch, there are some concepts to keep in mind. When possible, use stuff from your own yard rather than buying something from outside. This saves money and is much better for the environment than mulches sold in plastic bags that are shipped using fossil fuels.

Fall leaves are great, usually free, mulches. You might want to shred them with a lawnmower first, as shredded leaves are less likely to blow around and create heavy, overlapping layers that restrict air and water movement. If you don't have enough leaves in your yard, your neighbors will often be happy to give you theirs.

Clippings from mowing the lawn can be used as mulch, but should not be used alone, as they will pack too tightly and can actually stop rain from soaking through, turning anaerobic and smelly if left sitting too wet. Mixing them 50/50 with another mulch like your carbon-rich fall leaves works great to prevent that problem.

If you do have to bring in mulch from outside your garden, be careful that you don't import weeds. Hay is a bad mulch, as it's full of weed seeds, while straw should be nearly weed free.

Shredded leaves make a great weed-blocking mulch around edible plants. Plus, they're free!

WEED CONTROL

Weeds are the problem that takes down most new gardens and gardeners, and one of the biggest mistakes I made early in my time gardening in this way was not planning well enough to deal with weeds. Weeds might seem like a small issue at first, but they can quickly become overwhelming if left alone too long.

The good news is that, with the right information and planning, you can keep weed problems to a minimum. My instructor and mentor would always frame it as "The best defense is offense," where you avoid long-term, repetitive and reactive tasks like weeding. Reactive maintenance means you haven't put a strategic system in place yet!

Weeds are the biggest problem, and the most critical to control, in a new garden space. New beds may have lingering perennial weeds you failed to kill during soil prep, and the soil itself will be full of dormant weed seeds waiting for the right moment to germinate. As we've seen, planting and bed prep disturbs the soil, bringing dormant seeds to the soil surface, and there are often a lot of empty spaces in a new garden for weeds to set up shop and take over.

All this to say that new gardens need extensive, regular weeding. If you put in the time early on and have a strategy in place, the number of weeds will drop off rapidly as the garden establishes, bare spaces get covered, the soil is disturbed less, and you finish killing off any lingering perennial weeds. This is one reason you should consider establishing new garden areas one at a time rather than all at once, so you aren't hit by heavy weed pressure from all sides.

I recommend weeding your garden once a week during the growing season. That may seem like a lot, but if you go through weekly, you'll just be pulling small, newly germinated weeds and catching the first new growth of any lingering perennial. These weeding sessions will be quick and easy, and you'll likely catch most weeds before they become a problem.

Wait a few weeks—especially in warm, rainy weather—and the weeds will get huge, hard to pull, and weeding will suddenly be an unpleasant chore rather than a quick walk through the garden.

Being able to identify your weeds can help you focus your efforts. It can tell you how hard you have to work to pull them and how to prioritize your time weeding.

Keep weeds out of perennial crops, such as this asparagus, early and often. Once they've taken over, it's extremely hard to get them back under control.

Hand tools are great for removing weeds. A weed knife known as a hori hori is one of my favorites.

This gardener has surrounded their lettuce plants with an edible "weed" (chickweed) to reduce the presence of other, non-edible weeds.

Perennial weeds, like dandelions (edible!) or bindweed, will come back from their roots if you just pull off the leaves. Removing them means digging out *all* of their underground parts, which can be quite a bit of work.

An annual weed, like chickweed (also edible!), however, is easily killed by pulling off what you see above the ground. The key here is to get to your weeds before they produce seeds. With their fast lifecycle, annual weeds can quickly produce thousands of new seeds if you don't catch them young. Focus on pulling annual weeds before they flower and set seeds.

When in doubt, try to get all the roots of the plant, and certainly try to get them when they're young. A newly germinated dandelion can be removed with a quick pinch of two fingers, while a mature one will require a trowel or digging fork and some elbow grease to get it out of the ground.

Maintaining your mulch layer is your friend when combating weeds. Thick mulch will smother small weeds, prevent most weed seeds from germinating, and can even smother out some of the less aggressive perennial species.

If weeding has *really* gotten away from you and a bed has been taken over by aggressive, perennial weeds, often the best course of action is to start over. Dig out the plants you want to save—being sure to remove all the weeds from their root balls—and use one of the methods from the site preparation section given in chapter 2 to kill everything in the bed before replanting. Then make sure to put in the regular time weeding to keep it from getting bad again.

EVALUATING PLANTS

Though you do your best to understand your site and research the plants you're putting in your garden, not everything you plant will thrive where you placed it. Regularly going through the garden and taking stock of what's doing well and what isn't is a key part to maintaining a garden.

Now and then you'll get disease in your plantings. In the extreme cases, these will kill plants, which can be devastating. When that happens, though, it's clear what to do: Dig out the dead plant, dispose of it—not in your compost—and then plant something different in that space. Putting the same plant in the same space is asking for the same problem to happen again.

Less clear, sometimes, are the various leaf spots, mildews, and rusts that can disfigure leaves but will rarely fully kill a plant. In these cases, there are a few options. If the plant is still productive and attractive, you can just ignore it. A little powdery mildew, for example, is often more of an aesthetic problem than a serious one, so maybe plant something else to block your view of the damaged leaves and just live with it.

If the problem is more serious, impacting your ability to enjoy the look of a plant or harvest food from it, then you have other options. One is to try and reduce the disease pressure by changing conditions. This will depend on the disease, but often sunnier, drier, more open spaces with air movement will result in fewer foliage diseases. You can also look for more disease-resistant varieties of the same plant. Or you can try to find a way to treat the disease, but unless there's something special and irreplaceable about that plant, I don't recommend this—the fewer chemical inputs to your garden, the more sustainable, ecologically sound, and safe to eat it will be overall.

EVALUATING CONDITIONS

The amount of shade in your garden will change over time, and sometimes plants just get put in the wrong spot. Check in regularly and see if any of your plants should be moved to areas with more or less sun.

Plants that are getting too much sun will show burning on their leaves: brown edges, or even bleached patches on the leaves themselves. If light intensity is too high, this can be mitigated, to some degree, with extra water, but usually the best solution is to move the plant where it'll get a little more shade. Be that, if you suddenly lose your shade—say if a tree comes down—or if you move a plant from a shaded spot to a sunny one, the existing leaves may burn, because plants produce leaves adapted to the conditions in which they grow. Even sun-loving plants like cactus or agave will burn if moved directly from shade into full sun.

Wait until new growth emerges before deciding if a plant needs more shade. If the new growth is still burning, it's time for a move or to plant something nearby that will give it a little shade.

Too much shade is a much more common problem in most gardens, and will show itself a few different ways. The first things that will decline in shade are the flowers. They take energy to produce and don't give any energy back to the plant, so when light levels are low, plants tend to bloom less and make more leaves to maximize what sunlight they can. This may be okay if you're growing a plant primarily for its leaves—who cares if your tea shrub isn't flowering very much? But if you want flowers for their beauty or for the edible fruit they'll develop into, moving a plant into a situation with more sun is always the first thing to try if you aren't getting enough flowers.

Plants in shade will tend to stretch tall and loose, with more stems between the leaves, and they'll be more likely to flop over. Overly shaded plants will also grow and bulk up more slowly than a plant that's receiving enough sunlight. This may not be a problem, depending on what you want from a plant.

Take a good look at your plants. If they aren't doing what you need aesthetically or not producing enough food for you to harvest, consider moving them to a sunnier spot, or pruning back whatever tree or shrub is above them to let in more light.

When considering how a plant is or isn't performing, it's always helpful to consider its natural environment. If, for example, your rosemary is looking pale and sickly in a moist, shaded spot, remember that it's native to sunny sites in the Mediterranean, so try moving it somewhere sunny, warm, and dry.

Even shady places in the landscape can support a plethora of edible plants. Here beneath a pear tree, the gardener is growing broccoli, chard, strawberries, and chervil.

REMOVING PROBLEMS

Sometimes plants just have to go. Many beginner gardeners feel like a failure if they have to dig out a plant, so they keep nursing along something that isn't thriving, or struggling to save a plant that they don't even like.

I'm here to tell you that life is too short to grow a plant you don't love, or one that just doesn't perform well in your conditions. Get the shovel, dig it out. You can offer it to a friend if it makes you feel better, but don't be afraid to edit out plants that aren't working.

We've seen some of the reasons why you would remove a plant: You don't have enough sun, your soils aren't right, the plant's been hit by too much disease, you don't like the way it tastes, or the plant is fine, but you don't *love* it. Once those plants are gone, get excited for all that space you've cleared out to put in plants that you *do* love!

Spend time to evaluate regularly, looking over your garden at every season, and be honest with how a plant makes you feel, if you like how it looks, if you're actually harvesting what you thought you would from it.

REALIZING YOUR DREAM LAYERED GARDEN

Keep changing out plants for ones you like better and you'll end up with the layered garden of your dreams: full of beautiful, thriving plants that make you happy and taste delicious, require minimal work, and live in a flourishing community that is as beautiful as it is productive.

You'll see bees and butterflies, songbirds and amphibians, all enjoying the space as much as you do. That's when you know you've created a layered edible garden that will carry on producing and making the world a better place for the rest of your life—and your children's lives and their children's children's lives.

It's how all food gardeners should garden. Let the journey begin.

Investing in building your dream layered garden is well worth the time and energy required. You'll be rewarded with delicious harvests for many years to come!

RESOURCES

Plants

One Green World
This Portland nursery offers an amazing range of unique edible plants. Find detailed, useful plant profiles: www.onegreenworld.com

Phoenix Perennials
A nursery near Vancouver, BC, in Canada that provides retail and mail order options. Explore their supply of neat, rare edible plants as well as an exhaustive online plant encyclopedia: www.phoenixperennials.com

Whiffletree Farm and Nursery
Another Canadian nursery, Whiffletree offers a wide selection of perennial plants, including a great selection of fruit trees and shrubs: www.whiffletreefarmandnursery.ca

Hardy Fruit Tree Nursery
Find cold-hardy (down to zone 1) fruit trees and shrubs available from this Canadian nursery: www.hardyfruittrees.ca

Seeds

Adaptive Seeds
www.adaptiveseeds.com

Canadian and regional seed retailers:

Salt Spring Seeds
(annual food crops, herbs, and medicinal plants): www.saltspringseeds.com

Small Island Seed Co.
(specializes in rare, cold-hardy, perennial food crops): www.smallislandseedco.com

BC Eco Seed Co-op
(a cooperative of over twenty farmers growing ecological and organic seed): www.bcecoseedcoop.com

Annapolis Seeds
(a selection of perennial plant seeds): www.annapolisseeds.com

Gardening Advice

Local State Extensions
We don't have these in Canada, but
I often refer to Oregon State University:
https://landscapeplants.oregonstate.edu/

Epic Gardening Blog
www.epicgardening.com

West Coast Seeds
Articles and instructions on how to grow a wide
range of plants for a number of different regions.
Includes regional growing charts and information
on pests and diseases and organic growing practices:
www.westcoastseeds.com/pages/articles-instructions

Regional (Pacific Northwest resources I use):

Great Plant Picks
Comprehensive profiles for plants adapted
for the maritime Pacific Northwest region:
www.greatplantpicks.org

GrowGreen Guide
Eco-friendly lawn and garden resources for the
Metro Vancouver region. Includes online plant-picking
tool and ornamental/edible garden designs:
www.growgreenguide.ca
*Includes three food forest designs I created for the site
(Food Forest: sun bed, Food Forest: shade bed,
Food Forest: garden bed)*

Book

Pemberton, Trevor, Chris Marsh, David Gearing,
Wendy Stayte, and Plants for a Future (Great Britain).
2021. *Plants for Your Food Forest: 500 Plants for
Temperate Food Forest and Permaculture Gardens.*
First ed. Leeds England: Plants for a future.

ABOUT THE AUTHOR

CHRISTINA CHUNG is an enthusiastic gardener and educator based in Vancouver, Canada. She has been intrigued by the act of growing food since the age of six, when she discovered bok choy thriving under the partial shade of a hydrangea in her family's garden.

With a background in horticulture from University of British Columbia Botanical Garden's Horticulture Training Program, Christina has served as the program coordinator and as an instructor. She has designed and taught small-scale urban food production courses, as well as delivered practical, hands-on landscape and garden management training to students.

Christina's fascination with unique perennial edible plants has led her to explore the possibility of introducing thoughtfully designed multipurpose plantings into residential and urban spaces.

When she's not experimenting in her garden and greenhouse with her son, Christina can be found leading gardening workshops and sharing her passion for learning about plants from around the world through her work as Fluent Garden.

PHOTO CREDITS

Alamy: pages 12, 19, 25, 27, 38, 42, 55, 60, 74, 89, 102, 116 (top), 136 (top), 147, 176, 179 (top right), 190 (bottom right), 192

Andrea Jones/Garden Exposures: pages 44, 65, 72, 76, 79, 87, 127, 209

Christina Chung: pages 7, 9, 10, 22, 28, 29, 34, 37, 56, 66, 68, 80 (top), 81, 116 (bottom), 117 (top left), 121 (left, center), 123, 129 (top left), 130, 131 (bottom left), 132, 133 (left, center), 135 (bottom left), 137, 140 (left, center), 141, 142 (left, right), 143 (top left), 144 (top left), 151 (top left), 152 (bottom left), 153 (top, bottom left), 154 (left), 159 (top, bottom right), 162, 166 (left column), 174, 175, 177 (left column), 178 (left), 181 (bottom row), 183 (top left), 187, 188 (left, center), 189, 204 (bottom), 210, 214

Google Earth: page 41

JLY Gardens: back cover and pages 17 (top), 26, 30, 31, 46, 53, 63, 69, 71, 92, 119, 120 (top row), 121 (right), 134 (bottom), 138, 139, 142 (center), 143 (top right), 144 (top right, bottom), 145 (center), 150 (bottom right), 151 (top right), 152 (bottom right), 153 (bottom right), 154 (right), 155 (bottom right), 156 (bottom left), 157, 158 (top, bottom right), 160 (left), 163, 164, 165, 166 (right), 167 (bottom right), 168, 169, 170, 171, 172, 173 (top right, bottom row), 179 (top left), 184, 185 (top, bottom right), 186, 188 (right), 196, 202, 203, 205, 207 (left)

Joseph Tychonievich: pages 129 (top right), 140 (right), 146 (top, bottom left), 152 (top), 155 (top, bottom left), 158 (bottom left), 161 (top left), 179 (bottom), 183 (bottom), 185 (bottom left), 190 (top)

judywhite/GardenPhotos.com: pages 14, 32

Shutterstock: front cover and pages 4, 6, 15, 17 (bottom), 18, 20, 21, 24, 35, 48, 49, 50, 51, 52, 58, 59, 61, 62, 75, 80 (bottom), 82, 84–86, 88, 90, 91, 93, 94, 96, 97, 98, 100, 104, 106, 108–114, 115, 117 (top right and bottom), 118, 120 (bottom), 122, 124–126, 128, 129 (bottom), 131 (top, bottom right), 133 (right), 134 (left, top), 135 (top, bottom right), 136 (bottom row), 143 (bottom), 145 (left, right), 146 (bottom right), 149, 150 (top, bottom left), 151 (bottom), 156 (top, bottom right), 159 (bottom left), 160 (center, right), 161 (top right, bottom), 167 (top, bottom left), 173 (top left), 177 (right), 178 (right column), 180, 181 (top), 182, 183 (top right), 190 (left), 191, 194, 197, 200, 206, 207 (right), 211, 216–217

INDEX